The Emergence of Awe

The Emergence of Awe

Reimagining Human Progress for the 21st Century

Hayk Sarkissian

First published in Great Britain in 2025 by
Tandem Publishing Ltd.

ISBN: 978-1-0687613-5-5

10 9 8 7 6 5 4 3 2 1

A CIP catalogue record for this book is available from the British Library.

Printed and bound in Great Britain by CPI Group (UK) Ltd, Croydon CR0 4YY.

For my father and mother

Contents

7

Conclusion

The Same Kind of Thinking

We cannot solve our problems with the same kind of thinking we used when we created them.
— Albert Einstein

… where you began was in fact the end of who you once were?

*

Who am I?

A question you have undoubtedly asked yourself countless times.

What is my purpose in this world?

A question that has haunted every human for as long as we can remember.

What universal laws have shaped humanity's progress and identity throughout history?

Are they relevant for this new age of connectivity, uncertainty, complexity and unprecedented technological progress?

These are the questions we should be asking today.

*

Writing from a Birmingham, Alabama jail on 16 April 1963, Doctor Martin Luther King described the difference between just laws and unjust laws. He noted how a person has a moral responsibility to disobey unjust laws and agreed with St Augustine that 'an unjust law is no law at all'.

If laws are tools of a higher authority to guide human behaviour, how does one know the difference between a just law and an unjust law?

According to Dr King: 'A just law is a man-made code that squares with the moral law or the law of God. An unjust law is a code that is out of harmony with the moral law. To put it in the terms of St Thomas Aquinas: An unjust law is a human law that is not rooted in eternal law and natural law. Any law that uplifts human personality is just. Any law that degrades human personality is unjust.'[i]

By describing laws this way, he segregates them into two types of laws – the laws of man and the laws of a higher authority – to show that a balance between the two is where justice lies. He believed that a truly just world would exist only when human values are born from natural and eternal laws.

So, we must ask – how do we determine what is a higher authority and so a natural and eternal law?

Humans have always sought higher authorities to govern their ways of life – God, religion, ideology, political -isms, or a powerful individual. These have changed over time and across geographies, customs and cultures. Yet the higher authorities chosen by man have always been subjective. The laws of man have equally been derived from this subjective authority.

To fulfil Dr King's dream, humanity must elevate the laws of a higher authority to the only objective truth in the universe that is natural and eternal – that of objective reality.

How can we free ourselves from our human biases and understand objective reality when our comprehension of it is still so restricted by our abilities? Our sensory systems, our brains and our consciousness reveal only that which we can see, hear, touch, smell, taste and analyse with our information processing systems, in other words our brains. The full spectrum of objective reality is so broad that we may never truly understand it. That's why we have placed so many subjective ideals in its place over the course of our human history. To explain the unexplainable.

Furthermore, Dr King saw first-hand how the laws of man are themselves inherently archaic and rooted to the biases of previous generations. He lived in a world of racial and class warfare, a world on the brink of nuclear destruction between two great powers, as well as proxy wars fought in far-flung corners of the world. All in the name of freedom, communism, fascism or other man-made constructs of a higher authority. We thought we had solved these problems with the thirty years of relative peace that followed the collapse of the Soviet Union, the alleged victory of the liberal democratic model and the integration of the world economy.

However, still stuck with unjust laws, humanity faces the same challenges as before, amplified by the unheralded level of connectivity and complexity of the 21st century. Two great powers are on a collision course, localised wars are reigniting around the world, social and political extremism is dividing societies again – all exponentially enhanced by the power of our unprecedented technological advancements. The imminent potential of artificial intelligence combined with quantum computing, nanotechnology and bio-engineering will redefine what it means to be a human faster than we anticipate.

This inevitable symbiosis between humans and machines will enable ultra-fast problem-solving (which would have normally taken hundreds of years), intelligent automation and self-improving systems (to do our manual works, repair our bodies and enhance our minds on-the-go), and precise control over matter at the atomic level (to crash through scientific, medical and other bottlenecks). We will soon have the opportunity to vastly improve our understanding of objective reality. At the same time innovation in our political and social institutions remains static – we are creating new problems but using the old tools and unjust laws to solve them.

This critical dislocation between exponential technological innovation and static socio-political innovation will simply enhance our problems a hundredfold. As Einstein foresaw, the same problems will emerge if we do not change our ways of thinking.

This book is dedicated to Einstein's call to arms to change our modes of thinking, and to King's model of human progress through just and eternal laws.

Until his last breath, Einstein pursued a single, holistic and eternal law that would explain universal truths, whilst King fought for just laws for man until the moment his breath was taken from him. This book is dedicated to their pursuits and is an attempt to find harmony between them. It is the first book in a series introducing a new philosophy to address their warnings, focused on changing our perspective and reassessing how we think about who we are, our purpose in this world, and the hidden engines which have empowered and hindered our progress thus far.

Changing our perspective means changing the deep-rooted fundamentals we have held sacred for so long about ourselves, our societies and our traditions. Ultimately, this is what philosophy is all about – to push for change and establish a new paradigm of thinking to progress the cause of human existence. By making thoughtful changes in how we think about existence, our prejudices, our reality, and our daily lives we can align more closely to universal truths that are objective.

True harmony and justice, as Dr King envisioned, will be achieved when the values of Humanity align with the values of Objective Reality.

Change of perspective is not easy. It is an arduous philosophical revolution to align the laws of man with the laws of a different kind of objective higher authority. To do that we must explore who we really are, and what a truly objective higher authority can be, pulling back the veils that have blinded us for so long.

The time for a new way of thinking has arrived.

A new philosophy fit for the 21st century and a re-imagination of human progress.

That is why we must start with a conclusion.

A conclusion to who you once were.

1

The Emergence of Awe

Entropy and Decoherence

Introduce a little anarchy, upset the established order, and everything becomes chaos.
I'm an agent of chaos. Oh, and you know the thing about chaos? It's fair!
— *The Joker*

Joy had taken over every inch of my body. Not the fleeting feeling of happiness that momentarily passes through us, never embedding itself. This was something different, this was a joy of unity, of symbiosis. Something more eternal, and transformative.

I could not stop laughing. Although nothing was funny. This was simply the easiest way to express my emotions. An honest form of peace. A feeling of pure ecstasy – the embodiment of the word eureka. Yet, at the same time, nothing had changed. I was doing the same thing I had been doing every morning for the past four weeks. That day, however, my hands flowed effortlessly and smoothly across the keyboard as if possessed by the ghost of a great classical pianist playing his last encore to an endless audience. My mind had become an open vessel to a force I could not comprehend, filling me with ideas to my brim, until those thoughts naturally spilled out onto my hands. Hands which I had no real control of anymore. Thinking, feeling and moving became one and the same – an effortless stream of consciousness – a connection with something meaningful. The glistening sun blinded me as it pierced through the branches of the tree outside, as if toying with me, acknowledging this new harmony between man and the universe.

See, it would say to me, using its sign language of light, see how it feels to be me. The universe's manifestation of love and life piercing through one of its infinite conduits in the sky. The sun, our Sun, played with its gentle warmth across my face and shoulders, through the dancing leaves in the wind, allowing my eyes finally to see. See what we see. What we know. Who we are, when we let go. When we allow ourselves to step outside of being human and experience the beauty of our true, objective reality.

Can you feel it? Can you see?

Yes, I could.

Yet, I had no need for eyes.

It was a definitive morning in April of 2020. The days were no longer separated into weekdays or weekends, nor any working hours. Time just flowed. The pandemic had turned the world on its head. I had no obligations to attend physical meetings or social occasions or travel halfway across the world for work. My hectic schedule spent in planes, rushing from factories to meeting rooms to airports, had suddenly been filled with an unprescribed and enforced peace and quiet. I was extremely lucky that I could feel relieved in such a strange time, when so many others were suffering. I would wake up with a sense of guilt each morning that there was nothing I could do to help, other than make sure that the people who worked for me or my companies did not lose their jobs, and we could support the communities in which we operated. As governments around the world restricted travel and contact, the most impact I had was from a laptop and a telephone.

So, I made a pact with myself – I would not let this free time go to waste. For the previous decade or so I had devoured nearly every popular science book and scientific journal that I could get my hands on. Whenever I had free time, exploring the microscopic and macroscopic universes was my first port of call. I probably did this to satisfy my deep-rooted regret of never having studied physics at university. Instead, I had chosen to study history then law (albeit I read more philosophy, business and political theory books in those four years than historical ones). Perhaps I chose history as a rebellion against my scientist and mathematician father, who unconsciously tried nudging me towards his own path. I, on the other hand, had a desperate need for non-conformity. This rebellious streak proved useful in the end, as it gave me a broader view of the world from a young age; one which continued into adulthood. My early passion for history and philosophy, combined with my later hunger for science and business, neatly complemented my experiences in law, as my first career. This was combined with a unique insight into politics by way of osmosis, having grown up with parents who spent most of their adult lives in this field.

Back in the pandemic, I decided to do what was burning within me for years. I wrote.

I felt like my heart and my mind were exploding with ideas. Being a ferocious and wide reader, I had started to see patterns and links across philosophies, politics, economics, the sciences, and even religion and spirituality. I had something to say. Every morning, I would write for at least three to four hours, and after two months I had put together a bucket of brain vomit which could be generously called a manuscript. This manuscript was my manifesto on how to optimise human progress by redesigning our institutions to be more relevant to an age ruled by the re-emergence of populist and extremist politics, exponential change, instant communication systems, and intelligence technologies. It was an attempt to understand why, throughout history, humanity keeps falling into the same trap and cannot break away from its adherence to atrophied social norms, political extremes and negative economic externalities of progress. But who cared what I had to say, I thought.

I left that manuscript alone, and normal life eventually resumed. That was until I returned for a walk by that tree outside my home in late November 2024, the tree that so gallantly allowed me to peek into a world of *connectivity* that a human brain so effortlessly and naturally resists. A connectivity that had been lost by humans over millennia through the power of evolution, whose sole aim is to make sure that we survive, and in no way to reveal the truth of objective reality. A connectivity we had shunned by forgetting the flow of time and blindly following generational traditions that may no longer be relevant for this new world.

I stood in front of it, and I felt ashamed. Had I betrayed our pact?

At that very moment I was inspired to turn my manuscript into something more professional and accessible. Being a very private person, I also knew what I had to do went against my instinct. In a world where the truth no longer holds any value and everyone overshares their every thought, putting my private discoveries and ideas out for everyone to see and criticize, or even worse to ignore, was daunting to say the least.

And yet, a hook had been lodged within me.

This was a debt I could not walk away from. When I can't make up my mind on something, I always try to project myself onto my deathbed and imagine – would I look back and regret that I did this or that I never even tried at all. It became an easy decision. At worst it will just get lost in the vacuum of information, opinions and selfies out there in the ether. At best, who knows, maybe it inspires or interests one single person to change the way they think, and that person goes on to impact a community in a positive way. Anything can happen from there. In nature and in human history, as our final chapter will reveal, isolated interactions can result in an emergence of beautiful things. Maybe this short book will light a spark in someone who has something more important to say or do in years to come.

It was the act of writing – the process and the enjoyment of expressing something out of nothing – that was my true motivation. The fleeting feeling of the universe touching me. Of stepping outside of time and living momentarily in pure *connectivity* – where there are no boundaries between the human, quantum and stellar entities – a state of boundless energy transference. Writing the first draft was like a release of the bottleneck of my soul. Sharing it with the world may dislodge the hook.

The brief feeling of awe on that sunny afternoon was a connection to something new but, at the same time, to something utmost ancient. That connectivity exists through all strands of human thought – philosophy, history, science, even our economic and political lives – yet we have developed societies and cultures in a way which segregated them into distinct concepts or schools of thought. This is especially and most critically true of science, philosophy and history. The patterns and symbiosis amongst these three unnaturally segregated practices can teach us more about who we are, if they are studied and applied as one. An amalgamation of learning can reveal more useful patterns than the segregated practice we have taught for so long in our schools. From this study, we can derive more useful economic, political and social models to serve our needs as a more complex civilisation. We can understand how to adjust our institutions and systems for a future that's looking increasingly uncertain, volatile and exciting.

This symbiosis is the thematic skeleton running through this book. It is a layman's exploration of the concepts of time, entropy, emergence, equilibrium, and our endless search for meaning. At the same time, it is a practical and philosophical overview of how we can improve our lives, be it social, political or economic. I hope to explain the importance of a shift of our perspective in how we can finally break away from the cultural and systematic indoctrination that we are born into. From the moment of our first breath, we are inadvertently subscribed to certain values and traditions placed upon us by our parents, our schools and our societies. In a way, we have no choice who we become – real change becomes harder to achieve with each passing year.

We are at an inflection point in history – it is the right moment to question everything, seek new patterns, reveal new truths, build new systems.

To begin, we must first look within.

My perspective shifted on that afternoon back in 2020; something within me changed. Whilst writing a chapter about the flow of time, I believe I finally felt its touch, whether by serendipity or by the grace of something I don't yet understand. I had read that athletes, musicians and artists can enter a state of flow when they reach a deep focus through engaging a meditative state of being whereby actions are automatised and natural.[1] They experience time dilation, and a gradual emergence of a feeling described as *awe*. Somehow, I managed to get through the back door and touch this state of awe myself. That split-second of eternity remains within me to this day – I have been

1 Read *Flow* by Mihaly Csikszentmihalyi (Harper, 1990).

chasing that feeling in everything that I do since then.

Where did it come from? How did I get there? How do I stay there? Thinking back, one attribute I could clearly use to describe that feeling of awe had something to do with time. It gave me the sense of being outside of time. A step outside the normal flow of life and existence – into a state of constant connectivity, where time does not exist.

We spend our lives restricted in three spatial dimensions and one of time. Perhaps this was the fifth dimension.

Perhaps this was a slither of objective reality.

I began to explore.

*

Whether it is hydrogen fusing into helium to create stars (and later other elements across our Periodic Table), or the Sun giving life on Earth through photons interacting with cells in plants, or even our hot teacup warming our hands: our world is in a constant flow of increasing disorder. Disorder simply means the lack of arrangement or predictability in things – gasses, energy, nature, humans – we all move in our distinct ways that increase disorder, otherwise we would be in a still, boring universe. Without it, there is no change, no progress, no creation, no life. This disorder is scientifically called *entropy*.

The term entropy is used in many fields, from chemistry, biology, information theory, weather systems to statistical probabilities, but it is in thermodynamics where it is most relevant. At its core, entropy describes how energy naturally spreads and disperses within a system. It reflects the tendency of energy to move from concentrated sources to more diffused and evenly distributed states. This energy transference is the process which generates life, through interaction and connectivity, when energy is transferred from the Sun to the Earth as it's absorbed by the atmosphere, then by trees and plants through photosynthesis. Finally, to us and other living creatures through the consumption of fruits, plants and other creatures.

We see entropy in the disorder of gases after the Big Bang, in single-celled organisms in water, or when you make scrambled eggs. However, entropy on its own leads to chaos and disorder, it is not productive. It requires an opposing influence to organise and bring it into a more structured state. This disorder is counteracted by forces that pull it into states of order: gravity pulls gases together to form stars and planets, evolution merges cells until the most adaptable survive, heat turns the viscous eggs into solid nutrition.

This opposing process to entropy is best described as *decoherence* – an agent of order that counterbalances chaos. If we look at our universe on a microscopic level – it is simply a swirling blur of quantum particles, electromagnetic waves and gravitational forces. Decoherence pulls this bedlam into something more comprehensible for us to experience on a more macroscopic scale. Decoherence keeps us from consciously experiencing the quantum world's pandemonium, instead organising it into the tangible reality of objects, things, planets and people we can interact with.

Quantum physics is a difficult subject to tread lightly around. Even at an early stage of its discovery, the concept of decoherence split the physics community in half when the godfathers of quantum physics, Niels Bohr and Werner Heisenberg, argued in 1920 that a quantum particle transitions into a classical state (i.e. decoheres) only through interaction with its environment, such as measuring devices, other molecules in the air or even with light. Decoherence, the transition from a quantum state to a classical state governed by the laws of classical physics (as described by Isaac Newton), is the process which enables us to experience the world as we do – a world of objects, forces and causal effects. It occurs through measurement or interaction. This is called the Copenhagen interpretation and is the main school of thought today. Although the Copenhagen interpretation opens a can of worms regarding what is meant by 'measurement' and 'interaction', as well as other stranger concepts like the 'wavefunction' of particles that determine where and how they decohere when measured, understanding it as a process of transition from a quantum to a classical state is sufficient for our purposes. It explains how the chaotic quantum world becomes an ordered and tangible 'classical' world for us to experience.

Ludwig Boltzmann introduced this idea of entropy in the late 19th century, using the example of mixing cream into coffee. High-entropy states (i.e. chaos and disorder), where the cream and coffee molecules are mixed together and scattered, are more common, because there are more possible ways the molecules can be arranged in the mixture. Low-entropy states, where the cream settles separately from the coffee, are rarer, because there are fewer ways the molecules can be arranged with the cream floating on top, given the constraints of the surface and the cup. Our senses help us 'decohere' this microscopic chaos into an ordered world we can understand. They are the instruments of interaction and measurement we have with objective reality surrounding us. Our senses are tools provided to us by evolution. Decoherence, in turn, provides us with the steady, practical picture of reality that allows us to make sense of everything. It makes interaction easier, by allowing you to see your potential mate more clearly, as well as to hear the tiger in the bush.

Evolution's only concern is to provide you with the tools to enhance your chances of survival. With that comes a lot of sacrifice, namely we lose our ability to experience the entropic world – true, objective reality. Decoherence is our *subjective human* reality – a world we can experience through the limitations of our senses. Entropy is the quantum world's reality – a world of endless possibilities. So, what are we missing out on?

To explore this question, we must understand how energy in the universe is used. This brings us to the Three Laws of Thermodynamics.

First Law: The energy of the universe is constant – it cannot be created or destroyed, it can only be transferred or converted from one form to another.

Second Law: The total entropy of the universe is always increasing.

Third Law: The universe will never reach pure order and stillness.[2]

The First Law of Thermodynamics is fundamental to our existence. Energy is always conserved within our universe – it cannot be created or destroyed, only

2 To be more precise, the Third Law of Thermodynamics suggests that as temperature approaches absolute zero, the entropy of a system approaches a minimum, and the system reaches a state of maximum order, but it will never reach pure zero.

transformed. The universe cannot generate energy out of thin air; it must be transformed from something else. The world's most famous equation, formulated by Albert Einstein, of $E=mc^2$ is the perfect embodiment of this principle, showing that mass (m) and energy (E) are essentially two different forms of the exact same thing. This equation – used originally to explain nuclear fusion and fission – tells us that even when something is at rest, its mass has intrinsic energy. When this object is in motion, it gains additional kinetic energy (the speed of light squared, c^2), but the total energy – resting energy and kinetic energy – remains conserved. Put another way, Einstein showed that you, me, planets, stars are all connected.

The Second Law states that in the universe, entropy (chaos and disorder) never decreases. An omelette does not become an egg, and a broken glass does not fix itself. If you're building a home – to achieve it, you need a lot of disorder and chaos. You need to break the earth for rocks, cement and other materials (including to create fuel to transport everything). You need to chop down trees and cut them for your nice furniture, maybe kill a few animals for the leather chairs and for the food you need to keep you doing all this manual labour. It may take two or three years of blood, sweat and tears from all involved but the act of creation brings with it a lot of disorder and destruction. All this chaos will eventually decohere in a beautiful home – a new state of order, a fleeting state of awe. However, the house will eventually start to wear down and you'll need to fix and upgrade things. Until eventually, maybe once that home is in the hands of your great grandchildren or other people, it will be demolished. The analogy of the house is applicable to our planet, and our universe. Entropy will always have the last say.

The most common way to describe entropy is using the example of the movement of heat – in any scenario, whether opening the door of a sauna, putting hot water in a bath, or watching your teacup cool – this is entropy in action, heat moving from hot areas to colder ones. This occurs because there are far more ways for energy to be distributed evenly (high disorder) than concentrated in one place (low disorder). Entropy is a measure of this disorder – the number of microscopic ways a system can be arranged while retaining the same classical structure on a larger scale. The Second Law of Thermodynamics tells us that entropy always tends to increase, which is why heat flows in this way and why systems naturally evolve towards greater disorder over time.

Heat occurs simply when atoms are vibrating, increasing their kinetic energy to interact more. As they collide, they transfer energy which we perceive as heat. In a cold environment, atoms are relatively still and solid, with limited interaction: water can become perfect crystals. Atoms are only completely still as temperature approaches absolute zero (-273°C), but never quite reaching it. This is the Third Law of Thermodynamics, and it means that matter and energy will always exist in the universe in some form of energy, motion or quantum fluctuations. Even in deep space where temperatures are extremely low, some form of entropy exists.

This is the spark of hope.

This is why you need heat (vibration and movement) to create life, to melt ice into water, and more heat to turn water into vapour. Order is still and solid, disorder is vibrant and alive. Both thermodynamics and quantum entropy explain how uncertainty can result in beauty.

It results in the emergence of interesting things.

This is why we *need* chaos, as an agent of change, hope and life.

As the Joker says right before he blows up a hospital in the *Dark Knight* film to shut down the entire city of Gotham – it's fair. It will always result in a new state of order – but perhaps not the one you would like.

Now, how does all this chaos, disorder and decoherence relate to the state of flow and time, and eventually why does it matter to us?

Why is it important to think about these things when we have so many more important worries in this world?

And how does *awe* emerge from all this?

2

THE DECOHERENCE OF TIME

Flow and $E=mc^2$

We shall not cease from exploration,
And the end of all our exploring
Will be to arrive where we started
And know the place for the first time.
— T. S. Eliot

If there were no end, would our lives have meaning? I kept asking myself this question as we made our way back from the cemetery. It was the first time I had visited my grandfather's grave. He had died long before I was born. His tombstone, a silhouette of his profile made from a single ominous obsidian, towered over us as I watched my parents bow their heads solemnly for what felt like an eternity.

My mind was racing when it should have been quiet. If we lived forever, would we fight wars for land that was never ours, making claims of ownership with new lines in the sand? Would we create art, music and dances to celebrate our brief existence? Would we pray to the gods to give us passage into their afterlife? Would we think more about the future than about our needs today?

What kind of world would we create if time just … flowed?

As I got home and lay on the sofa quietly, my mother's soft humming soothed the bite of the cold breeze that would occasionally escape through the creaking windows and walls. Despite its drawbacks, the house had a soul. Perhaps it was my grandfather's soul, who had built it with his bare hands. A talented architect responsible for many of the buildings circling the Republic Square in central Yerevan, as well as other landmarks. He died at the tender age of forty, a life half-lived. We never met, but I developed a fascination with him. Maybe because my father and aunt said I had his character, or maybe just because we shared the same nose. Whatever it was, that imperfect home built by my grandfather felt like the safest place in the world, because each tile and panel was built to secure a loving embrace around his family when he was gone.

To this day, his memory is the only thing that would bring pain to my father's stern eyes, who lost him when he was only ten years old himself. A young boy who had to grow up to be the man of the house at such an innocent age. A young boy with a lost childhood who took the responsibility of his mother and sister on his shoulders.

Looking through my grandfather's photographs and hearing stories about him, I could tell he was an erudite man, a man of ambition. But he was held back by the system he was born in. Unfulfilled dreams, unsaid words, frustrated aspirations and shallow regrets are the roots that take hold within and grow into a cancer, whisking away our loved ones before their time. A man like him could have built some of the most beautiful buildings in the world, had he been given the chance. Instead, the closed Soviet system would allow him to achieve only that which was in their plan. A system built on restrictions – never allowed to connect, to interact, to destroy, to create, to achieve, to think freely. A closed system high on decoherence and low on entropy. Eventually, these restrictions turned into a cancer that took hold of the Soviet system's tentacles from within and led to its eventual demise. A decohered system, be it political, social, economic or scientific, can never create beautiful things. History has shown this time and time again, and even today not many people would choose to live in North Korea (possibly the furthest on the political decoherence scale).

Yet over time, entropy will always prevail.

The Laws of Thermodynamics harness the powers of interaction and connectivity to produce useful outcomes, such as life. Plants use the energy provided by sunlight to turn water and carbon dioxide into carbohydrates, trapping some of the original solar energy while releasing oxygen. Animals can gain this solar energy by consuming the carbohydrates (in the form of vegetables, plants and fruits, or each other), allowing them to live and grow. By combining these carbohydrates with the oxygen they breathe, animals release carbon dioxide and water back into the atmosphere, until the moment they cease to exist and turn into carbons themselves. Energy in the universe is preserved as the First Law of Thermodynamics dictates; it simply changes from one state to another. The Second Law reveals that at each step, a little disorder, or entropy, increases. Disorder results in life, powered by the energy transference contained in the rays of our Sun.[3] This relationship between energy, entropy and various states of order is what we have come to understand as *time* itself.

In fact, to truly understand this scale of our

3 For more details read Paul Sen, *Einstein's Fridge: The Science of Fire, Ice and the Universe* (William Collins, 2021) p. 123–4.

relationship within time, we must zoom further into the microscopic world of atoms and explore things at a Planck scale. Simply put, Planck scale is the lowest denomination of scale in physics – a boundary where classical physics breaks down and quantum effects rule. The Planck scale can be applied to space, energy (heat) and even time, because just as we cannot experience the atoms in our fingertips passing heat to the atoms within the pages of a book, a laptop or a tablet, we cannot experience a nano-second of time.

Time is not ubiquitous because it is experienced differently in different locations and by different observers or interactors. Special relativity says that the faster you go, relative to someone else, the slower time passes for you than for her. This phenomenon is called time dilation and becomes more pronounced at speeds approaching the speed of light. In general relativity, Einstein further discovered that the stronger the gravitational field (or the deeper you are in a gravitational core), the slower time passes. This means that time runs slower closer to massive objects, such as planets or black holes, compared to areas farther from such masses. That's why within a black hole, which is the highest concentration of gravity known to man where even light cannot escape, it is assumed time runs extremely slowly (if at all). These are Einstein's groundbreaking discoveries to show us that time is not a fixed or homogeneous experience but one that is *relative*, dependent on speed and gravity. He also discovered that time and space are completely intertwined into something called spacetime, but we won't dive down that rabbit hole just yet.

We can, however, explore why the experience of time can be just as distinct as a book, a chair or other objects we sense with our brains. We may see or experience all these things in the same way but derive completely different conclusions – you liked the book, but I hated it, the chair is comfortable for me but not for you, time passes slower for me but faster for you. This is because every creature in this world, including every person within humanity, has its own framework of senses with which it interacts with objective reality.

Our experience of the world relies on us breaking things down into structures. We have our house, our car, our job, our spouse, our chair and our computer. Our experiences rely on our ability to decohere the entropic nature of reality. This decoherence is a momentary experience of a ceaseless process of change. In that moment we can sit on the chair inside the house while typing an email on the computer. But even the chair, the house and the computer experience a process of change. Nothing is forever. If left over time, depending on the number and type of interactions with that object, they will eventually decompose. The chair will break if you sit on it for too long and your keyboard will be so worn out you won't see the letters on the buttons anymore.

Humans, however, prefer to create structures (consciously and subconsciously) in order not to be lost in a vast ocean of entropy. We need to create structures and logic that are grounded in stability for us to feasibly process this overabundant information flow. These structures are mostly stories that we have all agreed exist – nations, laws, companies, cultures, religions, traditions – if there was no tacit agreement between people that these things existed, they simply would not. Our evolution demands that we use language and stories to create these structures and boundaries so that we can live together more effectively as a species. This is why we are so successful at survival (for now).

One of the most critical things we erected structures and boundaries around is time itself. We have divided time into years then months then weeks and days, hours, minutes and so forth. However, this is a relatively recent innovation. For most of human existence, time was experienced as a continuous, cyclical flow. That was until the invention of clocks in the town square, which were eventually miniaturised into your pocket and onto your wrist. Even in the days of sundials and hourglasses, time was not kept but only its flow was measured. People lived by the flow of daylight, divided into watches by the Romans – not the physical things on your wrists but the guys who looked out for enemies in the distance. Initially watches were divided into two (before and after noon), then eventually into multiple watches in accordance with the sundial. Christian monks further decompressed this flow into a rigorous schedule of prayers in the latter half of the Middle Ages. Since 1967, the International System of Units has defined a second as the vibrations (or oscillations) of caesium atoms. Caesium-133 was chosen because it has a very consistent and measurable resonance frequency that is unaffected by most environmental factors.

Thus, through the advances of science, we have refined time from the flow of nature to become a decoherence of *quantum* nature. Step by step we have broken it down, given it more and more boundaries.

With this decoherence of time, we have shaped our lives around the vibration of a single atom. Clocks are intellectual technologies, initially used by elites, but once disseminated they established new modes of thinking, applied to everything we do. Time on a Planck scale, as it is defined today, is *an exact slice of entropy, now*.

Each tick of Planck time, which is approximately 5.39×10^{-44} seconds, represents an infinitesimally small moment in the universe's ongoing increase in entropy. Even at these minuscule timescales, the universe is never in a perfectly ordered state – there's always some degree of entropy at work. In fact, Planck time is so small that it defies our everyday understanding of time, emphasising how, even at the most fundamental level, the universe never achieves perfect order.

We consciously threw away the flow of time and replaced it with Decohered Time as our new master. Why did we do this?

Firstly, we experience the world in terms of things and objects, so it is a natural process for us to gradually decompress time into definitive objects, such as a specific when – or more specifically at eleven forty-eight in the morning, as I write these words. This anchoring helps us to anticipate the next flow of information and hence eases the burden on our processing system (our brains and our senses), allowing us to process more complex and useful information. Secondly, as we increased the number of tools we created from the earth and started to instil specialisation within our societies (farmers, builders, warriors, cooks, etc.), we needed a way to communicate and coordinate all these various strands. Better timekeeping made societies more efficient in specialisation and tool-making, and vice-versa. It became a self-reinforcing flywheel. Thirdly, we created this subjectivity of time because it is the nature of our information processing and predicting system (again, another way to say our brains, to be explored further in the fourth chapter). We needed a mechanism and a reference point to apply memory and foresight into our increasingly complex daily lives.

Our unique perspective of time is merely that, our perspective. Decoherence shifts us away from the microscopic, entropic interactions that are constantly occurring in the universe. Without our perception (and the process of decoherence through interaction), we would experience the universe as a continuous, interwoven collection of possibilities and states of connectivity. Instead, we experience it as a flow of sequential events because of how we are positioned in the universe, starting from an initial state of lower entropy. This initial low-entropy state shapes our experience of time as we perceive it: a forward-moving process. As the poet of time and famed physicist Carlo Rovelli explains, time is not an objective part of the universe but an *emergent* one, bringing us back to ourselves yet again:

'In every experience, we are situated within the world: within a mind, a brain, a position in space, a moment in time… In order to understand our experience of space, it is not enough to think of Newtonian space. We must remember that we see this space from inside it, that we are localised. In order to understand time, it is not enough to think of it from outside: it is necessary to understand that we, in every moment of our experience, are situated within time. We observe the universe from within it, interacting with a minuscule portion of the innumerable variables of the cosmos. What we see is a blurred image. This blurring suggests that the dynamic of the universe with which we interact is governed by entropy, which measures the amount of blurring. It measures something that relates to us more than to the cosmos.'[4]

Time, therefore, does not actually exist as an entity in and of itself. As Immanuel Kant presciently intuited, it is something that is projected out into the world by us. It is *emergent* from the relationship between entropy and decoherence. What this all means is that our perception of time emerges from the continuous flow of *events* – the shifts in entropy (increasing disorder) and decoherence (collapse of possibilities into order) happening around us. Every time there is a change in state, we interpret it as a moment of time passing. Therefore, time is not an independent entity but a byproduct of how we experience these changes in the universe. Perhaps now if we define time again as *an exact slice of entropy now*, it may make more sense.

To be more exact, time is a series of events or arrangements at a Planck scale that balances between the dichotomy of *entropic decoherence.* It is the natural phenomenon which enables us to feel this change, leaving traces of the entropy in our memories and an anticipation of future entropy in our predication machines. It is our own subjective construct. It is our justification which binds our experience of entropic

4 Carlo Rovelli, *The Order of Time* (Penguin, 2017) p. 134.

decoherence together in a more easily explainable way. It is, in the end, a useful story.

Entropic decoherence is a phenomenon which Max Tegmark, a famed physicist, argues helps our consciousness emerge:

'We don't know how our minds work, but we do know for sure that all information that ever reaches our mind from the outside world must first pass through neurons from our sensory organs, for example the optical nerves from our eyes and the cochlear nerves from our ears, which all decohere ridiculously fast. So, by the time we become subjectively aware of any observation about the outside world, things have already decohered, guaranteeing that we'll never perceive any quantum weirdness and explaining why we only perceive robust conventional states.'[5]

What Tegmark means here is that the quantum world of atoms, particles and energy transference is in constant chaos – a world of pure entropy. Entropy increases in isolation as things tend to become more disordered when they don't have external influences organising them. It is only when we engage with this world via our senses that this chaotic world decoheres into Newton's world of objects, causes and forces that we can measure, touch, smell, taste, see and

5 Max Tegmark, *Our Mathematical Universe: My Quest for the Ultimate Nature of Reality* (Penguin, 2014) pp. 211–22.

comprehend. This entropic decoherence happens so quickly we will never be aware of it. But it happens. It's happening when you read these words and when you blink or take a sip from your drink. Our mere interaction via our senses with the quantum world are the triggers of decoherence. Without this interaction there is no decoherence, and no world in which you and I can live. Only one of pure chaos.

Decoherence keeps us sitting at the edge of this chaos. Decoherence allows us to experience its beauty. We, as sensory and social beings, are decoherence manifest. Whereas objective reality is entropy manifest.

Our subjective reality, therefore, is relationship-based and not atomised or individualised, or simply there. It requires interaction and observation, to shift the perspective from a quantum world of chaotic entropy to a classical world of organised decoherence.

Entropy and decoherence together form our objective and subjective realities. The transition from one to the other spurs *change* – the true essence of time. At the same time (no pun intended), connection and interaction together spark the engine of life. Without the never-ending dance between entropy and decoherence, time would not exist, and we would simply not … be.

*

Our relationship with time can also explain many of the institutions we have created from within it. We naturally place this framework of time with a before and an after upon all our belief systems. For example, most monotheistic or polytheistic religions rely on a beginning. There must be an origin story or a way the universe was given birth to by a deity, following which all beings came to be via an act of creation. Similarly, there is usually an apocalyptic or celestial end, whether it is one's ascent to heaven, a higher level of being or a judgement day. These stories are all a part of the direction of entropy and our experience of change. We must have a beginning, therefore there must be an end, just as we experience birth and death. Such stories make most sense to us. Monotheistic religions adhere to the laws prescribed internally by the perception of time by our brains and our interactions with each other (in other words, our societies and cultures).

Notwithstanding this, there are of course religions such as Buddhism and Hinduism that reflect on our existence as an understanding without the restrictions of Decohered Time – the essence of universal oneness, or connectivity. Hinduism and Buddhism are philosophies based on escaping suffering by releasing oneself from the attachment to objects, emotions and people. Detachment, the separation from the material world, results in nirvana in Buddhism or in Hinduism the reuniting with Brahman, the universal soul of connectivity. The ultimate detachment is not really from people and objects, but of course, from time itself (and consequently from decoherence). Such religions uphold a perception of reality outside time as the zenith of their spiritual pursuits, unlike monotheistic religions, which exist within time. This is not a critique of one religion over another but an example of how our different cultures can also reflect different relationships with time and form an understanding of our reality.

Back in the real world, for as long as humanity has existed, time has generated a concept that has shaped our societies, politics, economics and our internal needs – that of *growth*. The concept of growth is a child of time because it is anticipating entropy and seeking decoherence. We seek change constantly to attain more – more money, more knowledge, more fitness, more experiences, etc. Having, needing, wanting and creating more is an intrinsic part of our experience of being human. It is what has enabled us to develop such complex societies and technologies so quickly. It is what got us to the moon and enabled us to harness the power of the atom.

The subjective experience of time has given us an innate need for growth – can we now use this gift in the right way? The immense technological, economic and social benefits of the last 200 years resulted from humanity's ability to develop this gift into a system called capitalism. Yet many negative externalities – wars, disease, poverty, inequality, climate crises – resulted from such unprecedented progress. This begs the question: is there a way to achieve progress whilst mitigating so many negative externalities? The answer lies within us, and how we perceive our place in this universe.

Would perceiving growth within a perspective of the constant *flow* of time, rather than as part of *decohered* time, improve our societies, politics and economics? Would we expend more effort on pursuing things that enable us to experience more connectivity and create value over a longer span of time rather than search for quick and immediate gains? Would it instil in us an urgency to create new incentives as individuals, corporations or nations? Would we design the same types of institutions that educate and employ people as we do now? Would we have the same foreign policies that can cause havoc for future generations and distant populations purely for the strategic and immediate gains of existing rulers? Would we implement business and economic metrics focused only on linear metrics such as GDP and shareholder returns? Would we make choices only for the benefit of ourselves?

The answers to all are clearly no. Time, it seems, is the ruler of man. Our paranoid and schizophrenic relationship with time – in which we need to break it into smaller units to derive more value from each unit, and then we put ourselves in a state of stress by chasing down each of those units to attain maximum productivity – is a testament to our greater sense of mortality. We know our time on this Earth is finite, so we must make every second count. I am as guilty of this as anyone. It is our nature.

Here we must stop and ask ourselves: how can we escape the trappings of Decohered Time and hence reshape our perception of reality to create better systems for a highly innovative, interactive and connected world? The answer has already been staring us in the face. It takes the form of equilibrium.

Let's take entropy as an example. If an open system continuously exchanges energy and matter with its surroundings, it can maintain low internal entropy. For example, living organisms are open systems that reduce their internal entropy by exporting entropy to their environment (e.g., through waste heat and metabolic byproducts), allowing them to maintain complexity and functionality. However, this process of exporting entropy does not negate the Second Law of Thermodynamics, as the total entropy of the system and its surroundings still increases. If entropy is too high without any decoherence, then this is pure chaos, a black hole where the normal laws of physics break down. In this state, there is a loss of meaningful connectivity or structure, and events become indistinguishable or absent. If decoherence is too high, then the system becomes completely and unimaginably dead to us. It is a place with no time, no space, no life. This is a place where there is no change in events or interaction or connectivity. A place of non-reality.

Our reality exists within the balance of this spectrum, this cosmic tug-of-war between entropy and decoherence. For us to perceive true reality we must find the point of pure equilibrium on the spectrum. It is at this point that we can experience the most productive outcomes from the harmonious dance of entropic decoherence. It is how the phenomenon of *emergence* comes about.

Emergence in physics occurs where complex behaviours or properties arise from the interactions of simpler components, that do not exist within the individual components themselves. It is a fundamental concept in understanding how macroscopic phenomena emerge out of microscopic interactions.

Wetness is a good example of emergence in physics. A single H_2O molecule is just a microscopic structure made of two hydrogen atoms bonded to one oxygen atom – it isn't 'wet' in any meaningful sense. Even a small cluster of water molecules doesn't display what we perceive as wetness. However, when billions upon billions of these molecules come together, their collective interaction creates the familiar property we know as wetness – the ability to stick to surfaces, form droplets and flow as a liquid. This emergence happens because of the hydrogen bonds between water molecules and their interaction with other materials, creating surface tension and adhesion that give water its characteristic 'wet' behaviour. This property then becomes useful, allowing for other emergent phenomena to take place such as the emergence of complex life within its environment.

The essence of emergence is that *the whole is greater than the sum of its parts* – new properties emerge at a higher level of organisation from the interaction of simpler components at the lower level. This is why studying complex systems often requires different approaches from studying their individual components – the emergent behaviours create new rules and patterns that can't be predicted just by understanding the basic parts.

The dimension of time, for example, is arguably emergent, as we have already discovered. To summarise, these theories in quantum mechanics suggest that at a Planck scale time does not exist as a fixed flowing entity but emerges from the interactions of the fundamental particles of our universe shaped by our senses and perceptions at the macroscopic scale in which we live. The concept of life itself, which emerged from 'non-living' interactions between single celled organisms into the biodiversity we see before us today, is the ultimate form of emergence we have evidence of today.

A state of equilibrium generates emergence.

It is where unimaginable possibilities from the quantum world can become beautiful things in our macroscopic world. This point of equilibrium gives us the ability to decohere our entropic reality enough for us to interact with it. This is where we can achieve a state of flow.

Ultimately, no matter how much we get distracted by other things in our brief lifetimes, pure equilibrium is the real search by humanity for its meaning. It is a human derivation of Nirvana and Brahma. It is our most optimal point of existence. Most importantly, it can be at this point of equilibrium where our pursuit of growth, knowledge and social interaction (including politics) can provide the most beneficial outcomes for humanity.

We can even find equilibrium in our daily lives when we talk about work-life balance, the never-ending tension between labour and capital, the competition between the political left and right, between capitalism and socialism, between the West and the Soviet Union, between entropy and decoherence. As with the spectrum of reality, we all know that either extreme of these dichotomies are not productive at best, and dangerous at worst. Equilibrium brings peace, productivity and progress.

We can dive even deeper within our minds in search of equilibrium.

Think for a moment about what meditation is. You sit still, close your eyes, focus inward on a mantra or your breath. After some time, you leave all your senses behind. Sight, smell, sound, touch, and finally Decohered Time. Only your energy remains, in search of pure connection with the universe, yet still within your physical self. A momentary state of equilibrium, connecting subjective reality and enlightenment.

The world's favourite equation, $E=mc^2$, is an embodiment of this search for connectivity. If we apply our high school maths to invert the equation by dividing both sides by the speed of light squared (c^2), the equation becomes $m=E/c^2$, showing that our mass (i.e. our physical self) is essentially energy divided by the speed of light squared. Meditation is an attempt to remove all motion and senses out of our existence to reach pure connectivity. If we were to collapse time or become timeless (i.e. $c=1$), then mass (you) becomes pure energy (when E is divided by 1, we get simply E), in a state of meditation. This is where we arrive at our flow state, where energy freely flows through us. This flow state is mathematically represented by Einstein's equation when we are most static, no longer bound by the restraints of time. A photon experiences no time passing because, from its 'perspective' (if indeed it has one), the moment of its emission and its absorption are the same instant. In meditation and a state of flow, people often describe a similar feeling of being deeply present, yet timeless, like everything is happening all at once yet slowly.

If we take this further and enter a state of deeper meditation, which only a handful of people can achieve by eradicating all senses and motion, we come to c=0. Using our hypothetical (and somewhat metaphysical) equation of $m=E/c^2$ as a reference to us as mass (m), then dividing any number by zero, results in *infinity*.

Perhaps this is the mathematical solution to enlightenment, where *E = infinity*. A dimension where we are fully detached from our subjective senses and Earthly reality, finally becoming one with everything else in the universe. Whilst this may sound esoteric or supernatural or like a shaman at Burning Man whispering sweet nothings into your ear – it is undeniable that the same equation we use to drive cars or fly planes can show us the truth about ourselves. We can indeed become pure energy and reach pure connectivity – scientifically and spiritually. We don't need to fear this but see it for exactly what it means – endless possibilities and opportunities about being human.

Buddhists call it Nirvana or the attainment of pure enlightenment, others may call it Heaven or Valhalla because it is where we find our purpose in between nothingness and pure existence. We all strive for it because it is our innate need to escape Decohered Time, a tool we have created ourselves to derive more units of productivity out of our reality.

It is where we finally find the emergence of awe.

*

So why is all this important? Why now?

Over the past century or so, perhaps we have swung too far into decoherence and tried to deconstruct everything in our lives, after having lived so many centuries in a more entropic world of high chaos, low connectivity and limited interaction (mostly via conflict). We have tried to measure and tame time, separating our reality into objects, actions and distinct outcomes, to an extent where we have forgotten the flow of time and how it helps us experience awe. Perhaps this change in perspective can inspire us or teach us how to reconnect with each other, with nature or with Objective Reality itself.

We are currently in a transitional phase moving from a classical Newtonian world to a quantum world, from a population boom to an older demographic, from a unipolar world to a multipolar world, from an Industrial Age to an Intelligence Age.

We have decohered time to its smallest degree and shaped our social, political and economic lives around it. Our political, education and healthcare systems have been shaped by outdated, simplistic, fragmented and reactionary approaches that are not as relevant in such an interconnected and dynamic world. These systems should now be transitioning towards more relevant, integrated and adaptive models that reflect the complex realities of an Intelligence Age.

As the world experiences a rate of technological innovation beyond Moore's law, fuelled by artificial intelligence (and soon, quantum computing), we remain rooted in the systems that were developed for the past. The path is laid out for us on a road of creative destruction, so it will never be a smooth journey, but if we remain rooted in the modes of thinking of yesterday, we will simply remain decohered to the past as the entropic nature of progress bulldozes our future. The frictions, conflicts and wars we see today in geopolitics, national politics, climate crises and many other issues stem from our references to time and growth.

We can choose to act on this by translating this pursuit of equilibrium into our systems – be they personal, economic, political or business-related – or we can continue to live through inertia as we always have done.

We should start with one small step.

With our own perspective of the world.

We need to experience time as it was intended – in a state of flow – and have the same mindset.

A ***flow mindset*** is always in search of equilibrium and balance. It is in search of pure connection with the universe. It is not thinking about just what's needed now and not always wanting more things. It is constantly learning, it is seeking wise and fair governance, it is striving for balance and searching for awe.

Understanding reality outside of time, and especially our own lifetime, is the only way to make political, social and economic changes to improve the path of human progress alongside our rapid technological progress. It requires a thoughtful and conscious change in perspective; a change in our modes of thinking, before it's too late.

Having a flow mindset, one that can always see things beyond the immediate and sits outside of time, is constantly learning and willing to fail, is constantly evolving and improving, is the only way to live in this Intelligence Age.

In a sense we are returning to where we originally started, as the words of T. S. Eliot describe. We have journeyed to the beginning of time, to the stars, to how we experience our reality. Yet we have come back only to find that true change starts from within.

It starts with you.

3

The Edge of Chaos

Connectivity and Creative Destruction

No man ever steps in the same river twice,
for it is not the same river and he is not
the same man.

— *Heraclitus*

The soldier looked at me, then at my hand, then back at me. Then smiled. I pushed myself up further on my toes, stretching my full five years as much as I could. He reached down and took the piece of torn bread loaf from my outstretched palm.

Spasibo, he said.

I looked up at my father seeking acknowledgement that I had not made any mistakes or done something wrong. He patted me on my back for reassurance and gave a resigned nod to the young soldier. We had queued around forty minutes that day, waiting for that loaf of bread. A ration from the government: good citizenry in exchange for a fresh piece of bread every other day. The great Soviet experiment of central planning had cumulated over seven decades into this pathetic state. But what did I know at that age, I was just happy to see a tank.

As we left the store, those steel behemoths were already parked in the street. Rugged Soviet soldiers poking out of their boreholes and standing at the sides of these menacing mechanical monsters, clutching death sticks in their hands and intimidating the gathering crowd merely by their presence. An air of confusion and unease permeated our neighbourhood. What were they doing here? Was I in trouble? I looked around and saw other neighbours in the street, quietly whispering to each other yet not daring to move. As my father held me closer to his waist, I realised that everyone was in trouble, not just me.

Recounting that day in 1989, when the existence of the Soviet Union was on its last legs, with protests erupting from Yerevan to Berlin, I recognised how quickly things can change. How much destruction had to take place for a new order to blossom from its ruins. I still recall the eyes of the soldier, perhaps only a decade or so older than me at that time. Fear, uncertainty, and confusion engulfed him. He was just a boy, born in a specific geographic location that determined the system in which he would grow up. He never had a choice. In a way his path was already set for him whilst he was still in his mother's womb. He would either hold that rifle or queue for bread with a ration token. The prejudices of generations before had been forced upon him from the moment of his first breath. Yet that instant, that gentle gesture I extended, gave him a moment of respite, as the world sat at the edge of chaos.

In the late 1980s, the Soviet Union was faltering – the economy was falling apart, dissidents were increasing, protests were erupting everywhere, and the United States was winning. Something had to change. Mikhail Gorbachev, the General Secretary of the Communist Party, introduced two policies as his Hail Mary last gasp roll of the dice. *Glasnost*, or transparency and openness, provided certain freedom to the press and admittance of past transgressions of the Communist Party. *Perestroika*, which loosened central control of the economy and businesses, introduced some foreign trade and eradicated subsidies, amongst other reforms. It was the financial and political restructuring aimed to save the Soviet economy from collapse – a dramatic shift along the spectrum of reality from a decohered society to a more open society. The door was ajar, and the Iron Curtain was lifted just a few inches. Chaotic entropy came pouring in. Soon, it became too late to stop the flood of change.

Whilst *glasnost* and *perestroika* were not the causes of the Soviet collapse, they were certainly the catalysts which accelerated its demise. They were supposed to be progressive and creative policies but instead resulted in the destruction of an empire. Creation led to destruction, and then finally to creation again as new states emerged from the rubble. One could not have occurred without the other. The destruction of an empire gave new states the ability to have a say in their own destiny. Yet this creative destruction started with a simple form of connectivity.

In fact, connectivity and interaction is how we came to be who we are, and I don't mean capitalists, but humans. It can be traced all the way back to the origin of the universe. If there had been no interaction among gases in the Big Bang, nor between single-celled organisms, there would be no life. From the first sign of the origin of the universe to the moment you are reading these words, everything can be simplified as a (very) long chain of interactions through the different phases of entropic-decoherence as we saw in the previous two chapters (i.e. what we call *time*).

All cellular interaction is determined by a cell's unique ability to extract energy from its environment and use it. Everything you see around you, as well as your body and your mind, is a process within a process.

These processes contain electrons jumping from one atom to another, creating a chain of reactions, all chemical, biological or quantum, at an atomic scale. Each interaction and exchange of electrons generates the next element on the Elementary Table. Combining these different elements in the right environment generates life. That's it. That's life. Just a bunch of electrons and elements constantly interacting with each other, from which complexity emerges. When the same interaction is viewed holistically, in a process of chaos-order-chaos, the subjective reality that you experience emerges.

For the first time in our human history, we are at a point where individual interaction between ourselves is almost instantaneous and physical reach is achievable anywhere in the world within hours. The power of physics has not only given us the knowledge to build aeroplanes and cars, but also to interact on a microscopic level through our quantum technologies such as smartphones and computers.

How did we get here? By exponentially increasing our interactions and hence our *connectivity*. The more connected we are, the faster the wheel of human progress spins – it is unarguable. In just fifty years, between 1950 and 2000, the world population more than doubled from 2.5 billion to over 6 billion, and by 2050 it is forecast to reach 9.7 billion people.[ii] Between the 18th and 20th centuries, it grew by less than one per cent. In just twenty years, between 2005 and 2024, internet access across the world jumped from 17 per cent to 68 per cent – further enhancing global connectivity. No wonder that the 20th century enjoyed the fastest growth ever, despite two world wars, the Spanish Flu pandemic and a state of constant conflict between two superpowers; all thanks to the power of what we call globalisation. Instead of direct competition through conflict and conquest, countries began to focus their resources on their strengths and trade amongst each other to make up for their weaknesses.

The 21st century's increasing virtual and digital connectivity will bring forth exponential innovation and progress, and at the same time the recipe for more destruction should the political systems governing this relationship remain the same as they were in the 19th and 20th centuries.[6] The era of free trade and globalisation did not last long though; coming out of the Covid-19 pandemic (itself a product of high connectivity), after a period of introverted policies to deal with the virus, countries began to revert to great power conflicts, xenophobia, expansionist agendas and zero-sum politics. Connectivity brings about both creative and destructive consequences.

This is not merely a human endeavour; it is prevalent throughout all forms of life. Recent studies show

6 To be explored further in the next chapter.

that plants and trees are connected by an underground network of roots and fungi – exchanging information and sustenance. They also grow upwards to reach for and be connected to the source of life, the Sun, to generate photosynthesis. This creates a natural circular harmony between the soil of the earth, the energy of the Sun and the water in the clouds. Together they combine in an elegantly connected system to create an ecosystem for life. We are also a part of this connected circular harmony. The tree that gives us the apple is largely made of carbon that must interact with sunlight and water to create it. We eat the apple to survive and eventually we exhale its carbon in a gaseous form which the tree needs to produce more apples. The cycle of life expands through various chains and connections of energy transfers, as the previous chapter revealed. It is a cycle full of destructive and creative moments – and the two intertwine like yin and yang to create the emergent phenomenon called *life*.

The animal kingdom also symbolises this connectivity through sustenance and procreation – the two inner motivators of existence of any living creature. Sustenance (a destructive action) and procreation (a creative action) connect them to their environment and necessitate certain cooperative and competitive behaviours within each species. This cooperative-competitive (or creative-destructive) tension is critical to progress and to life itself. The hunter-predator chain works across the whole animal kingdom (including humans) whether carnivorous or herbivorous. The purpose of this chain is the exchange of energy across different species to ensure survival.

We can source this chain all the way back to the plant kingdom which absorbs the photons from the Sun and starts the cycle of connectivity on Earth. This energy is then transferred across the food chain in the form of calories, from vegetation and fruits to grazing herbivores, to eventually the predators who hunt or farm them. The higher up the chain, the more likely you are to have access to the whole chain for more opportunities for energy consumption. The more collaborative a species, the more likely it is to control parts of the chain or the whole chain, as humans do. Some species even develop extreme collaborative relationships both as a part of their species group, such as a pack of hyenas or a pod of dolphins, or as a symbiosis between very distinct life forms, such as bees that pollinate flowers or coral reefs and the fish that populate them, turning reefs into the 'cities' of the sea. Inter- and intra-species symbiosis (an elegant form of interactive connectivity) can be found everywhere.

Humans, however, differ from other species in a particularly intimate way. Human babies need nurturing by their parents for several years before they even have a chance to survive. Most other species are independent from their mother either immediately or soon after the moment of birth. Most species are also used to the absentee father. A human baby is completely helpless as it cannot distinguish, speak or even stand until at least a year after birth and even then, is still highly dependent on its parents for survival. Chimps, for example, share the new-born among the females of the tribe, which consolidates their tribal values. Most mammals are on their feet at the time of birth and finding their own food within weeks or months as they adapt quickly to their environment. Comparatively, a human baby needs this connection for much longer to learn to survive in our complex societies.

This increased need for nurturing of our young arguably necessitated more interaction and collaboration, laying the foundations of what we call 'language', then 'societies' then 'culture' and finally the 'traditions' that tacitly dictate our way of life. These traditions become institutions under which we live and abide by today. Be it the concept of money, government or a corporation – these stories help us generate more platforms of connectivity. The higher the rate of connectivity and interaction, the faster the wheels of creative destruction turn, resulting in what we call *progress.*

*

Throughout history, the old have resisted the young, while the young have nudged the old, and so on and so forth. This tension has manifested itself between classes, genders, ethnicities, and age groups. The pre-Socratic Greek philosopher Heraclitus created the notion of *polemos pater panton* – war, or competition, is the father of all things. Competition is a process of destruction and is itself the spark of creation.

History has many examples of this duality, but no one exemplifies this on an innately human level better than a man called Fritz Haber. Haber was someone whom history can either describe as the saviour of millions of lives or as an angel of death who augmented the terrors and destruction of the world wars. In 1908, Haber invented a large-scale catalytic synthesis of ammonia from elemental hydrogen and nitrogen gas, reactants that are abundant and inexpensive in nature. He used exceptionally high temperatures and pressures to convert nitrogen into nitrate fertiliser in previously unimaginable quantities. Directly as a result of Haber's Nobel Prize-winning discovery, industrially produced nitrate fertilisers enabled the growth of infinitely more fertile crops across North America and Europe than anyone would have thought possible at that time. America's vast global wheat exports and the success of the EU's Common Agricultural Policy would not have been possible without the work of Haber. Neither would the transformation of Asia and the emergence of China and India as dominant 21st-century global economies. Norman Borlaug's miracle rice strains were grown with massive inputs of the nitrate fertiliser made from the process Fritz Haber discovered. His act of creativity fed the world for the next hundred years and counting.

At the same time, the 20 million deaths of World War I, the Russian Revolution, the rise of Adolf Hitler and the Holocaust may never have happened without Haber's contribution. When the British sea blockade cut Germany off from its vital raw materials at the beginning of World War I, it was the Jewish-born Haber who developed the techniques to produce nitrate explosives from ammonia that enabled Germany to fight on for another four years. His innovation was the driving force behind the 650,000 people who were killed in the Great War directly by the use of poison gas warfare. He personally supervised the very first gas attack in history against British troops at Ypres on 22 April 1915.

Even his innocent endeavours had catastrophic repercussions. He produced, almost as an afterthought of his poison gas research, a useful garden insecticide called hydrogen cyanide and then forgot about it. A quarter of a century later, this same insecticide, with its warning-indicator odour removed and relabelled Zyklon B, killed millions of innocent European Jews in the Nazi extermination camps. The children and grandchildren of Haber's sisters were among its many victims.

His work and legacy reflected the two-faced nature of modern science, providing unimagined advances in prosperity on the one hand, and inconceivable suffering on the other. In the Second World War, this was manifested in the form of the atomic bomb and the promise of endlessly clean nuclear power by J. Robert Oppenheimer, whereas today the same fear and anticipation circulates around artificial intelligence and its golden boy, Sam Altman of OpenAI. Fritz Haber is one of many embodiments of the creative-destructive property of nature in human form.

It has evolved over centuries as a parallel-process that is essential to all aspects of life – from the creation of life itself to corporate competition and innovation. It is the blood running through the veins of progress. It pops its head up everywhere we look. But where does this concept originate from?

*

Human life multiplies by creating new life and then is destroyed by nature, each other, or its inability to regenerate healthy cells. From this perspective, there is no purpose or emotion to life; it is just a pure process of multiplication through an expansionary process of creation and destruction. This is remarkably like another strand of species – the virus – which seeks to multiply in the host through a similar creative destruction process. On a global scale, humans embody the creative destructive process, and on a human scale a virus embodies it. Both are also portrayals of connectivity themselves.

Viruses predate humans, yet the increasing frequency and severity of their impact is a consequence of human interaction with each other and with the animal world – rapid population growth, increased domestication of animals, expansion of agriculture into wider ecosystems, urbanisation and industrialisation of food production – all enhanced it. Humans develop new industries, belief systems and technologies that create value to societies but also destroy existing elements; similarly, most people who encounter a virus will develop antibodies to become immune to the virus while others will suffer more and perhaps die. The destructive property of a virus is also perversely creative because it strengthens the human immune system.

Similarly, challenges that you experience or mistakes you have made will eventually enable you to form a resistance, and apply your flow mindset, growing at each point of destruction to a new point of creation. Within each destructive process lies the seeds for improvement. Destruction is not an absolute truth; it is more accurately understood as an integral step in the process of creation and innovation. Similarly, your failures, if framed correctly, are more powerful than your successes.

Now, since we are on the topic of small things that multiply, peering into our cellular structure, something very interesting happens here as well – our cells seem continually to deconstruct their own complex molecular structures and create new parts. It's as if our bodies are continually recycling themselves. Our cells contain enzymes called lysosomes and a protein called proteasome which work together to regularly create new molecules within our cells at a rapid rate, regenerating a new organ for you roughly every three days.[iii] As the pre-Socratic Greek philosopher Heraclitus observed long ago, the man changes as fast as the river he crosses.

When humans do not consume food for a long period (our sole source of energy), this activates our lysosomes to burn the less useful molecules to ensure survival and create only the most vital ones. Essentially your body eats itself to destroy the bad cells within you. This evolutionary quirk is present in the ancestors of plants, animals and humans, and is believed to have developed around two billion years ago. Today, this superpower has evolved into a dieting tool called intermittent fasting, whereby we limit the consumption of food to only a short window of time in the day to activate the lysosomes and recycle ourselves more optimally (and burn fat while doing it). Scientists are now studying how to engage lysosomes to help cure diseases like cancer and Alzheimer's by using the creative-destructive properties they have. They have noted that a possible reason these diseases arise with age is because age reduces the lysosomes in our bodies. Even on a cellular level, we cannot separate our natural propensity to create from the need to destroy. It is, in essence, how we move forward.

If you look across humanity's existence, you will see a history of different groups, in the form of empires or civilisations or nations, growing and then crumbling, the remnants of which come together and grow again from somewhere else or with redrawn lines on maps. Our cultural and civilisational history is a story of creation and destruction. The great Byzantine, Egyptian, Persian or Roman civilisations have not lasted until today, but their traces echo in our current civilisations. Even the act of us bearing children and then dying is the purest embodiment of this process. Nothing is forever and we are in perpetual flux.

Revisiting the birth of the universe and the concept of interaction brings us back to creative destruction, all the way to the first ever creative destruction. The Big Bang generated only the lightest elements (you can see these at the top of the periodic table) such as hydrogen, helium and lithium. When gravity brought these specks of dust together, it created a nuclear reaction, forming the first stars. The first stars burned their fuel quickly and were able to make only a few elements heavier than hydrogen and helium. Stars have since undergone a continuous nucleosynthesis process whereby they internally generate most of the key elements with which matter is formed. If they are big enough, they then undergo a supernova process in which they explode and expel all these elements across the galaxy, seeding the next generation of stars. This process has happened since the birth of the universe in several stages of the evolution of stars to create heavier and heavier elements. Stars are born and die but, in the process, they create the building blocks of life. This is the ultimate form of creative destruction.

If humans need to worship something, why don't we simply worship stars? They are the only real and objective evidence of gods we have found so far. Our creators are right there. We are made of their sacrifice. We harness their power to live, learn and create. They give us warmth, sustenance and life. The figure on the next page[iv] shows the elements present in a human body which are traced to the nucleosynthesis process. Take a moment to breathe in this fact, but the fact that we are made of dying stars is as undeniable as anything else we may want to believe about ourselves.

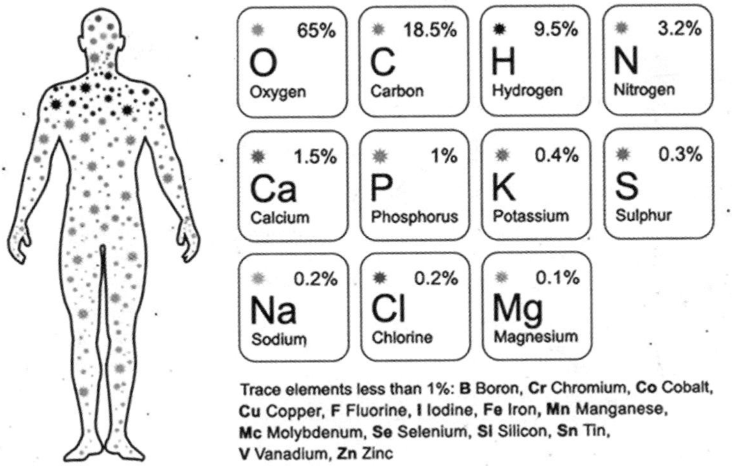

Are we made of stardust?

We must now pull ourselves back down to our own reality and deal with earthly matters to see how creative destruction applies to our everyday lives. In fact, you may have heard the phrase 'creative destruction' before: certainly, if you are involved in finance or economics. The term was coined by the Austrian economist Joseph Schumpeter in 1942 to explain how to increase productivity in the manufacturing process through innovation. He considered this the fulcrum of capitalist society – a 'process of industrial mutation that incessantly revolutionises the economic structure from within, incessantly destroying the old one, incessantly creating a new one'.[7]

If we were to simply remove the words 'industrial' and 'economic' from that famous phrase, let's see how it would read and if it could be applied to what we have been discussing before:

> *A process of mutation that incessantly revolutionises the structure from within, incessantly destroying the old one, incessantly creating a new one.*

It perfectly fits the description of the various types of creative destructions we have explored. This could very easily describe the process of lysosomes regenerating the cells in our body, as well as the nucleosynthesis taking place within stars to create elements of life, as well as our evolutionary path from our simian cousins to upright mammals. Not only is this process a part of our existence but it is something we have unwittingly integrated into our model of economic development and production, giving it many other names. From a practical human perspective, creative destruction is the process of innovation. It is what has made us accelerate through the stages of normal evolution and enter the realm of science fiction; things which were not

7 Joseph A. Schumpeter, *Capitalism, Socialism and Democracy* (Routledge, 1996 [1942]) p. 82.

possible before are now everyday occurrences thanks to this divine and inherent gift. Schumpeter merely pointed out what had been happening for centuries in the skies, in our bodies and in the way that we interact.

*

The last two hundred years have seen a dramatic acceleration in the pace of innovation. The 1800s were very much a different world to the one we live in today: 85 per cent of the world's population lived in extreme poverty; the average life expectancy was 31 years; 193 countries practised slavery and half of all children would die before their fifth birthday;[8] not to mention the lack of any effective medicine and the widespread beliefs that natural events were the wrath of God or the work of demons. During this same period, the progress of industry, business and technology is clear for all of us to experience today as we read books on our computers, order food on our phones, fly around the world in hours and get entertained in three clicks. We often forget that the destruction of the old businesses, technologies and industries are not only inevitable but also necessary to pave the way for these innovations.

The definition of progress includes the fact that the previous form – whether it is a species, business,

8 Hans Rosling, *Factfulness* (Spectre, 2018) pp. 52, 55, 60, 61.

scientific theory or structure of a belief system – will no longer survive in the changed environment. This is how evolution works; it is how stars gave birth to humans.

There is a certain order of things until a small mutation creates a new order. Laws must be broken or bent, and the boundaries must be pushed before the standards can be reset. The mutations that set or adapt best to the new order or laws are the survivors and victors, while the previous iterations will erode into the sands of time.

The natural process of progress requires this duality of creation and destruction to occur hand-in-hand, so we should not mourn the crumbling of our pillars of yesterday but try to understand how to mitigate the negative externalities caused by their downfall, while building stronger and more sturdy pillars from the rubble.

This chaotic stasis is the seat at the edge of chaos. It is a paradox. A state of stability brought forth only through constant change – and endless cycles of creation and destruction. This paradox is resolved only through the attainment of states of equilibrium.

The process of creative destruction is essential to human progress. But this does not mean we must have wars and conflicts to attain it. There is a place for us, just at the edge of chaos where progress can emerge without such negative externalities. This place is emergent from the balance achieved between creation and destruction, between connection and integration. Optimal progress is only emergent from this state of equilibrium.

To achieve this paradoxical state, we need to firstly acknowledge that we are shaped by prejudices beyond our control from the moment of our birth. Family, society, culture and institutions are all what make us who we are, but ultimately shield us from our true self, and from a reality which is not useful to understand for our survival. If we pull back the veils hiding such prejudices, which help us be so productive but at the same time keep us so blinded, we may discover a new system built on stronger foundations.

We need a new perspective that will make us finally change how we are living, learning and thinking together in ways that are not hamstrung by previous generations' biases and mistakes, disguised as traditions. We must finally move away from our collective solipsism and see the truth about ourselves.

We must pull back the veils of prejudice and discover who we really are.

1. Fractals: infinitely complex patterns created by repeating a simple process in an ongoing feedback loop. Nature is full of fractals (trees, rivers, coastlines, even broccoli), showcasing the beauty of entropic decoherence.

2. Batman and the Joker – a popular representation of chaos and order.

4. The changing density of a cloud of atoms cooled lower and lower (going from left to right), approaching absolute zero. The sharp peak in the later graphs confirms the formation of a Bose-Einstein condensate (BEC) – a fifth state of matter (after solid, liquid, gas and plasma). BEC consists of atoms so cold that they condense into a quantum state, acting as one in a state of minimum entropy.

3. *An Experiment on a Bird in the Air Pump* (1768) by Joseph Wright of Derby. His paintings depict early scientific experimentation related to thermodynamics, a concept that caught the world's imagination.

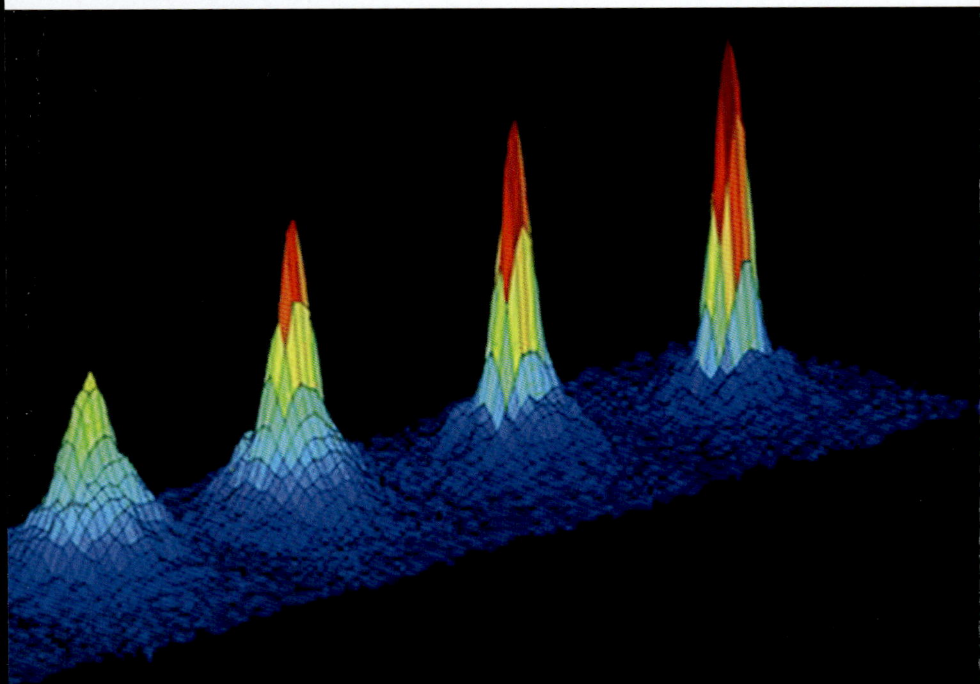

5. The Aztec Calendar Stone, also the 'Sun Stone', reflects Aztec perception of time as cyclical, rather than linear; repeating eras or 'suns' represent different epochs in the universe's history.

6. The Torre dell'Orologio in Venice displays the phases of the moon, highlighting its cyclical nature. Mechanisms show the position of other celestial bodies, linking earthly time to cosmic rhythms. This intricate design reflects a Renaissance fascination with astronomy and astrology, blending precise timekeeping with a broader understanding of the universe's influence on human life.

7. The Buddhist 'Wheel of Life,' or Bhavachakra, depicts samsara – the endless cycle of birth, death and rebirth, driven by karma and desire. The layers include the six realms of existence, the causes of suffering, and the forces that keep beings trapped in this cycle, such as ignorance, attachment and aversion.

8. Caesium atomic clock made by Louis Essen and J. V. L. Parry, 1955.

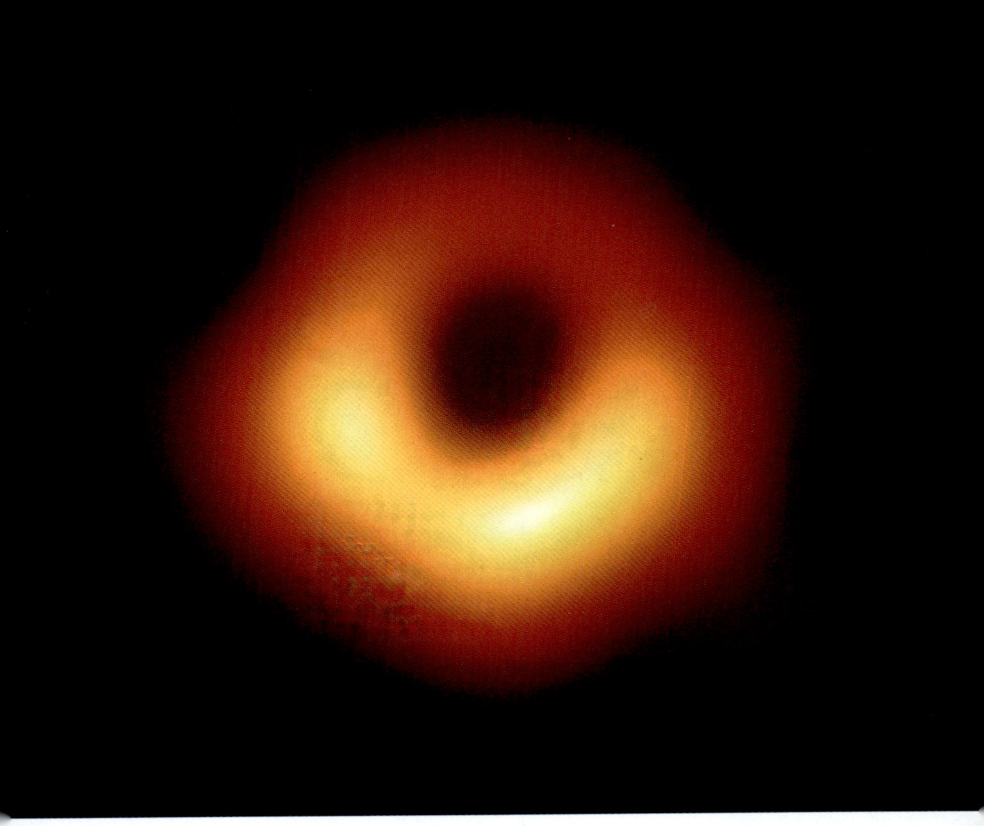

9. The first picture of a black hole. The event horizon – the point of no return – lies within the bright disk of material (accretion disk) swirling around the black hole before falling in.

10. All-sky map, showing microwave radiation released just 370,000 years after the Big Bang. Tiny fluctuations in the image eventually gave rise to the clusters of galaxies that populate the universe today.

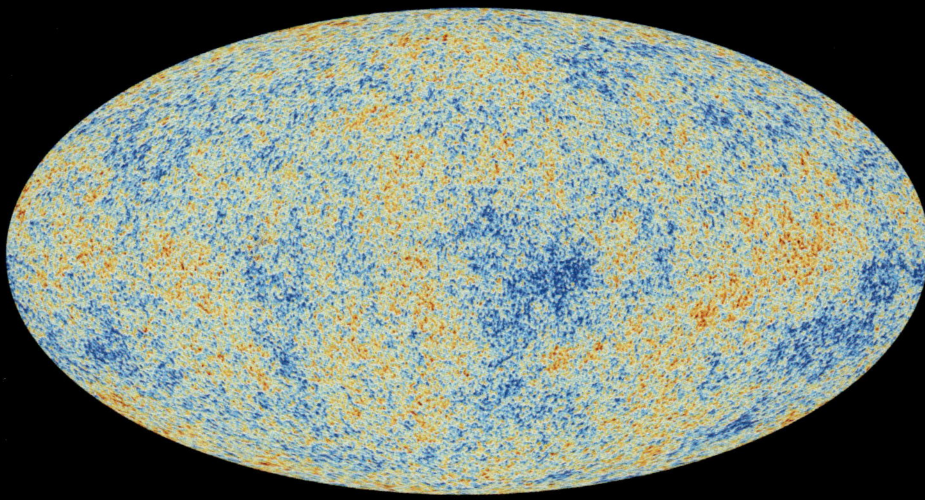

11. Cassiopeia: a supernova remnant. A dying star explodes as a supernova, scattering elements across space. This violent destruction seeds the cosmos, forging new stars, planets and eventually life: nature's ultimate creative destruction. Was the Big Bang also the death of a star … or something else entirely?

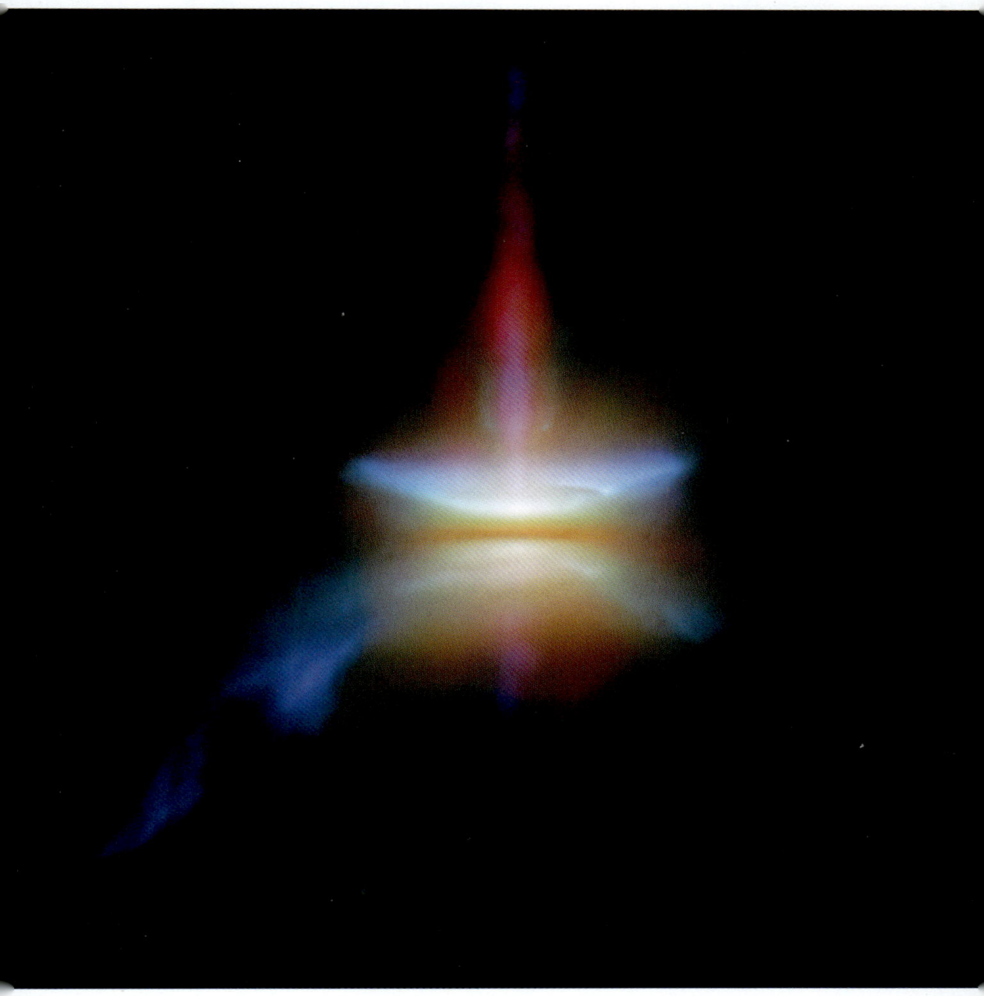

12. Birth of a star. A close-in image of a protoplanetary disc around a newly formed star.

13. *The Cave of Plato* by Jan Saenredam.

14. Three polyhedral by da Vinci, an in-depth study of the golden ratio, or 'divine proportion'.

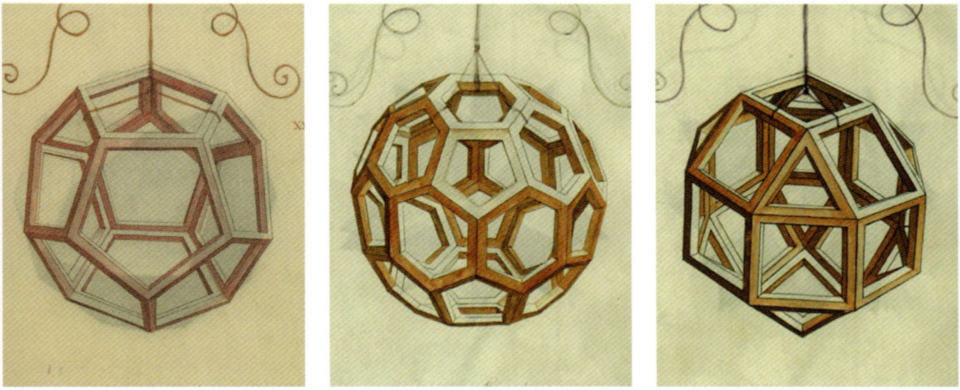

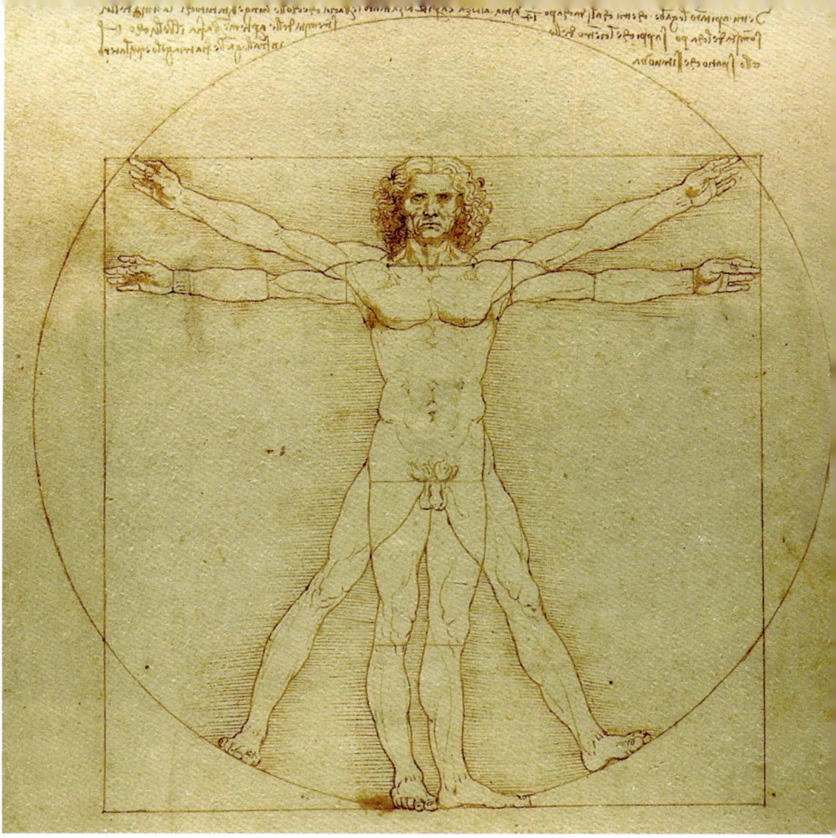

15. da Vinci's *Vitruvian Man* showcases the ideal human proportions, inspired by ancient Roman architect Vitruvius.

16. Albert Einstein's office mere hours after he died, Princeton, April 1955.

17. Geocentric v. heliocentric models of the universe. The geocentric (above) depicts Earth as the centre of the cosmos, a view held for centuries. The heliocentric model (below) places the Sun at the centre, revolutionising our understanding of astronomy.

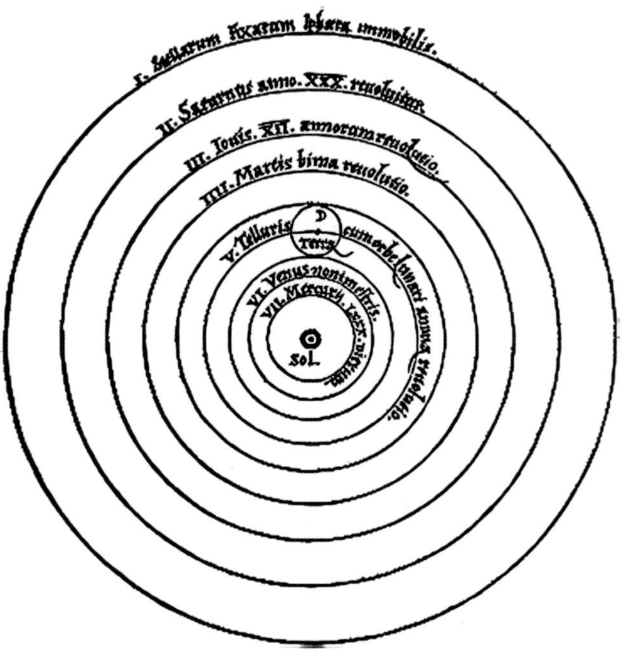

18. Nicolas-Sébastien Adam's *Prometheus* (1762, Louvre).

19. Model of a quantum computer interior (IBM Thomas J. Watson Research Center).

20. A famous oscillatory chemical reaction propagating across a petri dish, often referred to as the Belousov-Zhabotinsky (BZ) reaction, illustrating the principles proposed by Alan Turing in his 1952 paper 'The Chemical Basis of Morphogenesis'.

21. Human Brain Folding (gyrification). The way the human brain folds into its wrinkled structure follows mechanical buckling principles, a process of morphogenesis driven by growth constraints.

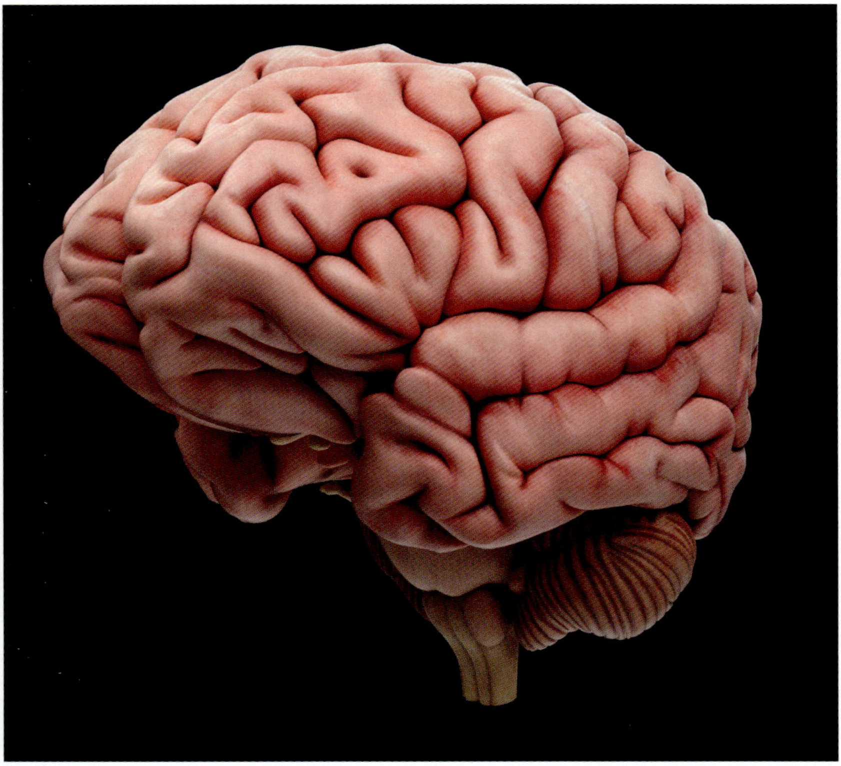

22. Lichtenberg figures emerge when electrical charges diffuse and branch through insulating materials, creating fractal-like pathways – like morphogenesis in nature, but with electricity instead of chemicals – a clear parallel with plant life reaching for the sun.

23. The three protagonists of *The Emergence of Awe*:
Alan Turing,
Albert Einstein,
Martin Luther King Jr.

4

The Veils of Prejudice

Consciousness and Equilibrium

The only real revolution is in the enlightenment of the mind and the improvement of character. The only real emancipation is individual, and the only real revolutionists are philosophers and saints.
— Will and Ariel Durant
The Lessons of History, *1968*

Nestled in my mother's warm embrace, I could not help but stare at the flame dancing gracefully at the candle's tip. As it swayed from side to side, I was entranced by its freedom and power – its ability to give us light and warmth, yet at the same time, a capacity for pain and destruction. I would frequently test these limits as a child by seeing who out of my brother and I could hold our hand to the flame the longest. Being the younger of the two, I would reconcile my regular defeats simply to inexperience.

At times our neighbour would come over and play the guitar for us, but tonight we had to make do with its echo from his house next door. That night we had to stay home. The flames of the candles dotted around the living room kept rising and falling in anticipation of an arrival.

Electricity was a luxury to us in Yerevan. On a good day, the electrons would flow through the house for at least four or five hours, but sometimes we would have to make do with just one or two. None of that mattered though, so long as we had each other, and so long as we could make each other laugh. I would always entertain the steady stream of family and friends that would come through our forever-open doors, organising the other children into different roles, with a full script and props. After just an hour of practice, we would showcase it to our parents, gathered around a table full of dried fruits, nuts and endless streams of coffee, waiting for their version of the cinema they could never have.

I loved stories as a boy, especially stories that would connect people who didn't know each other with hearty laughter or heated debate. As a child I felt my sole purpose in life was to make my mother laugh. Whether by singing, dancing, acting or just plain silliness, I would draw out her smile without fail. A smile that covered me in more warmth than a thousand candles ever could.

But that night, I did not have it in me to perform any acts. There was too much anticipation in the air. After months away, my father was returning and bringing with him what lay on the other side of the Iron Curtain. I could barely remember what he looked like: he had been away for over a year, but the images he brought with him from another world were imprinted in my mind since the first time I saw them.

In 1988, after the devastating earthquake that rocked Armenia, killing thousands and leaving many without a home, protests erupted around the nuclear power plant by a small group of individuals that feared an aftershock could cause a nuclear disaster. Their objections caught the mood of unease in the country, which gradually turned into hysteria.

The inevitable shut-down of the plant meant that the main energy source of the country was no longer operating, eventually resulting in electricity being rationed. The upside to this terrible situation was that, with no electricity and not much choice for entertainment, it allowed our imaginations to run wild. I always think that it was a blessing in disguise, as it instilled a creative spirit within me which I carried into adulthood.

I needed that imagination because the toys sold in

the Soviet Union were, well, Soviet. If you were lucky enough to have a television, you would quickly realise that it had a complete disregard of children. Except for one cartoon of a wolf endlessly chasing a rabbit whom he could never catch.

Nu, pogodi! the Russian wolf would always say at the end when the rabbit got away. *Well, just you wait!* We waited. Trapped. Every day, trying to get away, but the wolf would reappear. Soviet citizens waited a long time. But the promise was never fulfilled. Only the lie grew taller. And the wolf kept reappearing.

Luckily, my father was able to attain a prestigious visiting professorship at Cambridge University and earn more than other academics back home. This meant access to toys from the West! He was an astrophysicist and a mathematician who dabbled in complex computer algorithms (what we call artificial intelligence today). He was also obsessed with new technologies, so in due course we became the only family in the neighbourhood to own not only a television, but a Video Home System.

Each time he would return from England to see us, he would bring back two or three cassettes with pre-recorded children's programmes for me and my brother to watch over and over, until we had memorised every single line from every single cartoon. The colours, the creatures, the creativity! The audacity of their imagination! A whole new world seemed to exist out there that we were not aware of. These Transformers, Thundercats and Pepsi ads opened a door that could never be closed again. But for now, we could only pull back the veil slightly and peer through a keyhole.

In the 1960s our parents had the same experience through rock and pop music. Anyone who could get their hands on a rare Beatles vinyl became a neighbourhood hero. This indirect interaction with the West showed us that anything was possible: something the Soviet system went to great lengths to ensure you did not dare contemplate. In a communist country, your path was always limited, linear and with a low ceiling. So, all we had were these sounds and images, through which we felt we were interacting with a different world. A world full of creativity. A different universe in fact. Freedom was the word they used a lot, even in the cartoons. It showed us what freedom felt like, looked like – a magical place where *change* was possible – the opposite of the static Soviet state of life.

Many historians credit this tacit interaction with the West as a big factor in generating the will of the Soviet people to push for independence in their respective countries. Experiencing this phenomenon first hand, I can testify that it really did influence how we evaluated our place in this world and helped us ask the right questions. Helped us ask *why*. Interaction was the first step towards change. The closed door between two political systems now had a form of connection.

In this context, when two large population groups were isolated for so long, connection became the most valuable thing in our lives. How could we live in such isolation anymore?

Once you had looked through the keyhole, you could never look back. Interaction sparked the connection. Connection led to the destruction of an archaic system. The destruction set the foundations for the creation of an era of prosperity and possibilities. There was no turning back once we knew what lay behind the Iron Curtain. We had finally stepped outside of Plato's cave and seen the shadows for what they really were.

Plato used the allegory of a cave to go further than explaining social and political restrictions (Figure 13). If you imagine yourself chained within a cave your whole life and you only see the shadows of people and the objects they carry flickering against a wall lit by a fire; this is your perception of reality. A philosopher, or an explorer who tries to understand the world beyond the senses immediately available to her, can free herself from her chains and step outside of the cave to finally see the fire which casts the shadows. Whilst confused, this philosopher may venture out further and see people and the Sun, and a whole new world that seems unbelievable to someone who has lived in a cave for so long. When this freed prisoner goes back to release the rest of the prisoners, they think she has gone mad and prefer to stay in the cave. They don't believe her because it is easier to resist the truth, which is too hard to accept. Acceptance requires leaving behind all that they knew from the moment of their birth. It requires the destruction of all they once held sacred.

Plato felt that we are trapped in this cave and a philosopher's job is to break those chains. This is also what a true revolutionary is in the words of Will and Ariel Durant, whose quote at the beginning this chapter guides us. Will and Ariel Durant were American historians and philosophers, best known for *The Story of Civilization*, an eleven-volume history of the world. They viewed history as a holistic concept inseparable from other subjects. The Durants believed that real revolutionaries are those who, like Plato's philosopher, break free from the illusions imposed by society, education and tradition. Instead of blindly accepting what they are taught, they question, explore and seek deeper truths – challenging the status quo, escaping the cave, unafraid of what they may discover outside, even if it changes all the assumptions they held for so long. These philosophers are people who look within and understand that what they are taught by their families and societies are but mere shadows flickering against the cave wall.

This is how a true philosophical revolution begins.

From the enlightenment of the mind and the emancipation of the individual's character.

In other words, from within our own consciousness.

*

How our consciousness emerges is still unclear, despite all the scientific tools we have had at our disposal to understand it. There are many schools of thought, which are beyond the scope of this book, but we can start with our senses and see where it leads us.

Human beings are one of the most successful information-processing systems on the planet. This is because we evolved to have an increasingly efficient and useful process of collecting information from our environment via our sensors (eyes, ears, skin, nose, mouth), feeding it back to our core sensory software-processor (brain) to digest this information and then react to the environment (world) via our integrated mechanical hardware (bodies).

Our physical bodies are also a complex system that has, through evolution, attained an equilibrium amongst a myriad connection of specialised tools (organs, muscles, etc.) and communication systems (sensors, arteries, nervous system, etc.) from which a human emerges. The neurons in our brain are specialised cells, just like any other cells in our body. Yet these neurons allow transmission of electric signals to all other cells in response to the sensory inputs received by other cells around our body. Our cells are tools that are specialised in their function, but it is the balanced communication between them that gives rise to a coherent organism – us. Neuroscientist David Eagleman compares this process to an ant colony, whereby each ant behaves in a simple, localised way, reacting to chemical signals from its environment and its genetic coding, operating as an autonomous unit without any centralised decision-making. But together, the complexity of ant colonies displays extremely sophisticated behaviours akin to a super-organism that can accomplish extraordinary feats.

The same happens in our brain and when these processes are taken together and interacting in the right way, the mind and consciousness emerge.[9] We can even draw the same parallel of the ant colonies to human organisations – such as a state or a nation or a company – and any other complex systems.

However, how can consciousness, which feels internally emergent, occur from our senses, which are externally induced? The answer comes from a misunderstanding of how our senses process information. Perception, for example, is not a passive recording of sensory data – it is an active process where the brain constantly generates predictions about the world and refines them based on incoming information. As

9 David Eagleman, *The Brain: The Story of You* (Canongate, 2015) p. 193.

explained by consciousness expert Anil Seth, it is 'an inside-out neuronal fantasy that is reined in by reality, not a transparent window onto whatever that reality may be'.[10] Since sensory inputs are often noisy and incomplete, relying solely on raw data would be too overwhelming for our brains. Instead, the brain anticipates what it expects to perceive and adjusts only when reality deviates significantly from this expectation. This predictive approach makes perception more efficient, allowing the brain to focus on meaningful changes. For instance, when you reach for a coffee cup, your brain estimates its weight based on its past experiences. If the cup is unexpectedly light or heavy, the resulting prediction error prompts you to adjust your grip and refine future expectations. Or simply to drop the cup.

As Seth explains, there is a parallel process taking place between our brains' predictions of what our subjective reality should be in *each slice of entropy now*, and the information received through our senses from the objective reality around us. The brain uses this information to form predictions about the world, constructing a model of reality.

However, this information by itself is not enough to generate consciousness. Seth argues that *integration* also occurs in parallel, whereby the brain organises and connects different pieces of information to create

10 Anil Seth, *Being You* (Faber & Faber, 2021) p. 83.

a coherent, unified experience of the world. It is the process of combining sensory inputs from different senses (sight, sound, touch, etc.) and integrating them with past experiences to form an ongoing, stable model of reality. This is another duality at place that we have seen within Nature and objective reality so many times thus far. From this parallel process emerges our consciousness, which is an equilibrium between information and integration (or entropy and decoherence in our parlance).[11]

So, as two leading consciousness and neuroscience experts separately affirm, for either a complex living organism or a consciousness to emerge, a key element is needed – equilibrium. Equilibrium generates a state where all the individual parts of a system are functioning together in a balanced, coordinated way, enabling the system to operate effectively as a whole and which leads to complexity.

The individual units must find a harmonious balance to be productive. The same rule applies to cells to generate a human, as they do to humans that generate a nation or a company, or companies that generate an economy. In any complex system – whether it is the brain, a society, or even a company – the individual components (like neurons, people, or business units) must work together smoothly to function effectively. If

11 Ibid. p. 55.

any part falls out of sync or there's too much chaos or imbalance, the system will break down. Equilibrium generates highly productive and emergent phenomena yet again.

Self-organisation becomes impossible if entropy is too high because the system is too chaotic for stable structures to emerge, and too static if entropy is too low (i.e. high decoherence) because the system is too organised to have any meaningful change or respond productively to new conditions. For example, our bodies can sustain life only within a narrow temperature range, where the energy of molecular movement is balanced with the energy needed to limit disorder. This balance creates a stable environment at the molecular level, allowing cells and the body to function.[12] As these cells find equilibrium for our complex bodies to emerge, our neurons also self-organise for our consciousness to emerge – all driven by the energy that we draw from our surroundings. Similarly, if an economy or society is too entropic (no rules, no regulation, no institutions) it becomes anarchy. Too centrally organised and it becomes a totalitarian state.

The brain itself is an organ that acts as a self-organising processing unit – a central command system of our consciousness to help us interact with

12 Neil Theise, *Notes on Complexity* (Spiegel & Grau, 2023) p. 66.

the rest of reality. But what do we mean exactly by our consciousness?

To find out we should dig deeper into how our cells interact with each other. Our sensory organs feed information via photons, temperature, texture, pressure, molecular concentration, and many other ways. These sensors are translated into electrochemical signals and through neurotransmission are fed into our brains. The hundreds of billions of neurons in our brain interact through electric pulses with other neurons every second to create a perception of subjective reality. This is the most magical form of equilibrium of entropic decoherence.

Gradually, these perceptions result in physical changes in our brains as well as chemical changes. Eagleman demonstrates these physical changes through the example of London taxi drivers, who must memorise intricate streets by heart, thereby developing a visibly larger posterior hippocampus than control groups. Even Albert Einstein's brain, which was analysed in depth to understand the source of his genius, developed a giant fold in his cortex, not because he was so smart, but because of his affinity for playing the violin. Piano players, because they use both hands, develop the same physical changes in the brain but in both hemispheres. The process of creative destruction is also taking place within our brains as we interact with our environment positively and negatively every second, to create a cocktail of chemical signals shaping who we are as individuals.

This process, combining entropic decoherence with creative destruction, is essentially how our consciousness emerges. It is the way the processing of complex information feels. When we sleep, no new matter is created but the structure of the properties within our brain is rearranged, hence why we still sense things in our dreams but not in the same way when we are awake. The prediction machine is still working and generating senses from within, but this is not complemented by the interaction with photons and other atoms when our senses are 'awake'. The act of dreaming proves that we do not need light bouncing off objects and into our eyes to experience a form of reality. We are still conscious but in a different arrangement, and that consciousness has nothing to do with external stimuli.

Another example is when ice melts into water – the information is rearranged but the matter remains the same. Thus, changes in the arrangement of matter (whether it's the molecules in water or the connections between neurons in the brain) can lead to a different state or experience (liquid vs. solid, or waking vs. dreaming) while the fundamental matter of our brain – just like the molecules of H_2O – stays constant.

The key aspect for the process of *emergence* to take place is, again, the attainment of *equilibrium*.

And so, a truth is revealed about our reality and our universe:

Consciousness emerges from states of internal and external equilibria.

*

Evidence of other equilibria are everywhere. The mathematical constant pi (π) is the ratio of a circle's circumference to its diameter, approximately 3.142 if we stop at three decimal places, a number that creates the perfect shape – the circle. The beauty of pi is further reflected in its never-ending decimal expansion. Every sequence of numbers, from simple patterns to rare and random sequences, appears somewhere in pi's infinite string. This randomness represents an equilibrium between order and chaos, further symbolising how complexity arises from fundamental constants.

You may have also heard of the golden ratio: a number equalling 1.618 called phi (ϕ). It is associated with the Fibonacci sequence, a series whereby each number is added to its previous number in the sequence from 0, 1, 1, 2, 3, 5, 8, 13, 21, 34, 55, 89 and so on. This unique number comes from when you divide a Fibonacci number by the one before it, the ratio gets closer and closer to 1.618 – the Golden Ratio.

It has been used in countless other fields from when it first appeared in Euclid's *Elements* in 300 BC, a seminal work of mathematical geometry.

It first appeared in biology when the Italian

mathematician Luca Pacioli published *De Divina Proportione* in 1509, and was best captured by Leonardo da Vinci's illustration of the *Vitruvian Man* (Figure 15). It appears in flowers and seed heads, such as in sunflowers or pinecones, where the seeds are arranged in a spiral pattern. The number of spirals usually corresponds to consecutive numbers in the Fibonacci sequence. Shells, tree leaves, pinecones, snowflakes and in some cases parts of the human body grow in relation to this Golden Ratio – not because it is from a divine hand, but because beautiful complexity emerges from states of equilibrium.

The Golden Ratio isn't just a mystical number but illustrates how balance is central to the natural world. This idea of equilibrium is echoed in various philosophies as well. The Buddha taught the Middle Way, advocating for a balanced approach between extremes. Aristotle described the Golden Mean, where virtues lie between two vices, for example, courage is the mean between cowardice and recklessness.

Only through a state of equilibrium can the beauty and complexity of our world emerge. A world of awe.

Howard Marks, a famed investor, highlights in his book, *Mastering the Market Cycle*, how financial markets always swing from exuberance to dread, from bull to bear markets, but they always come back to a point of equilibrium where they are most productive.[13] He constantly emphasises that the market tends to gravitate back to the mean. The further it deviates from that mean, the stronger the pull to return to it, which is why we experience cycles of boom and bust. These cycles are essentially just the market moving away from and then correcting back towards equilibrium. The same can be said of politics – a strong lean to the left eventually results in a wave of right-wing governments gaining power as a reaction to the extreme left – as occurred in the 1930s with communism, and is happening again in the 2020s. We have had many times of extremism in our short history of humanity, but our long-term survival depends on us reverting to an equilibrium as a civilisation as often as possible.

States of equilibrium are not only sources of emergent phenomena of being human, but they can also be spotted everywhere in the universe. Earth is ideally located in the Goldilocks zone – just far enough from the Sun to allow for water to exist on the surface at perfect temperatures, but close enough for it not to freeze over. Gravity pulls on us just enough to halt any atmospheric desolation that would make Earth more like the Moon, but not too much to crush any hopes of life. Earth's orbit is extremely well-balanced

13 Howard Marks, *Mastering the Market Cycle* (John Murray Business, 2020).

without any larger objects disrupting the ease of its orbital rest. The delicate balance between photosynthesis and carbon production by plants has provided the circular economy of life that has given us other complex organisms.

Most importantly, equilibrium results in the complexity of our own brains, as neuroscientist David Eagleman explains. He describes two types of neurons: excitatory neurons, which stimulate neighbouring neurons, and inhibitory neurons, which prevent their neighbours from being overstimulated. These two types work together to keep the brain flexible and balanced. If there's too much inhibition, the system becomes rigid and can't change. If there's too little, neurons compete too much, and no clear outcome is achieved. A well-functioning system requires a balance between these forces, a Goldilocks zone where everything is balanced just right.[14]

Eagleman goes even further to suggest that all systems, including multicellular organisms and governments, thrive on this delicate balance – existing on the edge of chaos. The right amount of competition fosters progress, but too much can destroy the system. This equilibrium is key to keeping any system ready for change and growth.

The complexity of the brain emerges because

---

14 David Eagleman, *Livewired: The Inside Story of the Ever-Changing Brain* (Vintage Books, 2021) pp. 185, 190.

equilibrium allows the brain to adapt, integrate diverse inputs, and evolve. This dynamic balance between stability and adaptability enables the brain to process complex information and exhibit behaviours that go beyond simple animalistic responses. Equilibrium in this sense is not just about maintaining stability; it's about creating the conditions for greater complexity and sophistication.

Extremism is nowhere near as productive. A well-functioning system, whether in the brain or in governments, requires the right poise between these forces, a 'Goldilocks zone' where everything is balanced just right. Eagleman suggests that systems, including multicellular organisms and governments, thrive on this delicate balance – existing on the edge of chaos. The right amount of competition fosters progress, but too much can destroy the system. This equilibrium is key to keeping any system ready for change and growth.[15]

Sir Martin Rees, Britain's foremost astrophysicist and cosmologist, has captured these forms of equilibrium mathematically into six constants that determine the recipe for a balanced universe which can give rise to stars and life if they are tuned to the right ratio:

'Is this tuning just a brute fact, a coincidence? Or is it the providence of a benign Creator? I take the view that it is neither. An infinity of other universes

15 Ibid.

may well exist where the numbers are different. Most would be stillborn... We could only have emerged (and therefore we naturally now find ourselves) in a universe with a 'right' combination. This realisation offers a radically new perspective on our universe, on our place in it, and on the nature of physical laws.'[16]

Each of his six states of equilibrium has a unique role to play in our existence.

ε (Epsilon) = 0.007 is the nuclear efficiency constant that determines how much hydrogen fuses into helium in stars. This value allows stars to burn steadily for billions of years, providing stable energy for life to evolve. We would not exist if Epsilon was 0.006 or 0.008, because it defines how atomic nuclei bind together.

N = 10^{36} is the ratio of electromagnetism's strength to gravity's strength. This massive number ensures that stars can form, and that the universe can expand properly, while allowing planets to hold on to their atmospheres. It is the mathematical core of connectivity – any deviation from it would not allow for evolution or creative destruction to occur.

16 Martin Rees, *Just Six Numbers: The Deep Forces That Shape the Universe* (Weidenfeld & Nicolson, 2015) p. 4.

Ω (Omega) ≈ 1 is the density parameter of the universe. This describes how close our universe is to the critical density needed for a flat geometry. Its value near 1 allows for the formation of galaxies and large-scale structures. Omega is the ratio between gravity and energy expansion, which was tuned perfectly to prevent universal collapse immediately after the Big Bang but allowed for it to form at the right pace.

Λ (Lambda) is the cosmological constant, representing dark energy. This tiny but non-zero value controls the expansion of the universe, preventing it from either collapsing too quickly or expanding too rapidly for structures to form. The best current estimate of this number is Λ≈1.1×10^−52 m^−2.

Q ≈ 10^{-5} is the ratio of the universe's gravitational energy to its rest mass energy. This determines the size of cosmic structures and allows for the formation of galaxies and stars.

D = 3 The number of spatial dimensions in our universe. Having exactly three spatial dimensions allows for stable planetary orbits and complex chemistry – essential ingredients for life.

*

Thus, from equilibrium emerges our objective reality. So why do we experience only our limited subjective reality?

Three veils of prejudice keep us chained in Plato's cave.

We sit on Decohered Time's arrow (described in chapter 2) as it shapes our destiny, as well as our consciousness, casting the first veil of prejudice over us.

The second veil is just as powerful.

A trick that has been played with humanity for thousands of years. We are still watching the shadows on the cave wall, and we are even retreating further back into the cave. Even more worrying is the realisation that those shadows are not even people. They are puppets held up on a string.

Evolution, our clever puppet master, has been slowly deceiving us, pulling us further and further away from objective reality. It has empowered us with godlike abilities to extract technologies from the earth, yet at the same time pulled a veil over our senses to simplify our world for the sake of our survival.

Subjective consciousness developed because of our evolutionary trick of being able to process information in a more sophisticated manner. Our sensory world, and our perception of reality, has increased and been overloaded with more information, which is entropic in nature. Subsequently, our ability to process this information more efficiently is an act of decoherence, enabled by the power of evolution.

If consciousness is defined as a state of being aware of and responsive to one's surroundings, we should understand what we mean by our 'surroundings'. If it relates to the decohered world of objects and things then indeed we are very conscious; however, if it relates to true objective reality then it is a lie.

We thought we are highly conscious beings, but we never tried to understand what true consciousness means.

Let's start with the most basic sense we use – sight. The light we see is a small part of the total spectrum – we do not experience ultraviolet, infrared, microwaves, X-rays, cosmic rays, etc. – where some animals do. Matthew Bothwell, a popular scientist, uses the metaphor of a piano to describe how we only see one octave whereas the electromagnetic spectrum stretches to 65 octaves[17] – we have remained blind in an ocean of light. Birds navigate using magnetic fields, bats and whales use echolocation, some snakes see infrared and bees experience ultraviolet waves. Our closest companion, the dog, has roughly fifty times as many scent receptors as a human. We are living in a tiny fraction of the total spectrum of reality – what we can call *Objective Reality* – shaped by our prediction-machine brains and our highly limited senses.

Donald Hoffman's illuminating book *The Case*

17 Matthew Bothwell, *The Invisible Universe: Why There's More to Reality than Meets the Eye* (Oneworld, 2021), p. x.

Against Reality definitively argues how evolution has favoured *Fitness* (i.e. senses attuned to survival) over *Truth* (Objective Reality) using the analogy of a computer desktop as our subjective reality of space and time, which allows us to do useful things. The true reality of nodes, wires, electrons and complex connections lies behind the desktop screen, from which emerge a simplified interface to allow us to survive more efficiently. Take human vision, as he argues, which is shaped by natural selection over aeons and compressed into three dimensions of space and one of time, into objects, shapes and colours. This compression omits some information such as the muons that streak through our bodies each day, damaging it with ionized radiation, but we see enough to survive and create offspring:

'The eye sports 130 million photoreceptors, which collect billions of bits each second… The circuitry of the eye can, with little loss in quality, compress those billions of bits down to millions… It then streams the millions of bits to the brain through the optic nerve. This stream … is a flood which would overwhelm the visual system if untamed… Billions of bits enter the eye each second but only forty win the competition for attention.'[18]

18 Donald D. Hoffman, *The Case Against Reality: Why Evolution Hid the Truth from Our Eyes* (Penguin, 2019) p. 155.

Evolution has built a data compression algorithm within us.

It is such a clever trick that we feel as if we are advancing as a species faster than imaginable (and indeed within our subjective reality we are) but we are not truly conscious of our objective surroundings. The fact that we have no idea what 95 per cent of matter and energy composing the universe are – calling them Dark Matter and Dark Energy – gives us a hint of how well this compression algorithm has worked. Evolution, through the power of creative destruction, has been excellent at decohering Objective Reality into objects, things and a sense of Decohered Time, so that we interact more efficiently, and survive more easily.

That is its sole purpose – to ensure our survival. It gives with one hand and takes with the other. Evolution is about *being*. Objective Reality and hence true consciousness are about *knowing*. Therefore, are we truly conscious? Are we really awake?

The answer is that we are further away from this Objective Reality than our ancestors in loincloths were. Objective consciousness is about understanding the *truth* of our being and our unprejudiced surroundings, not restricted by our senses.

Thus, Evolution casts the second veil of prejudice over us – but the deeper message of all this is that our perceptions mislead us about the very nature of the universe. Decohered Time and Evolution are the first and second veils of prejudices that define who we are as humans, allowing such progress in our subjective existence and civilisation, but at the same time pulling us further away from understanding the truth of our existence.

There is a third and perhaps most important veil which also blinds us to Objective Reality – that is imposed on us by ourselves, and one which we can actually do something about.

From the moment of our birth, we are taught to live under the societal norms in which we are born. We are taught what is right and what is wrong – not only in how we act but in what we believe – by our parents, schools, friends, colleagues. As babies and children, we are told how to eat, talk, act and think. Generations of prejudices are passed on to us from the moment of our birth to help us survive in our complicated societies, which would ostracise us if we do not.

You are not who you truly are – you are a reflection of the time and place you are born in. A young boy born in Soviet Armenia in 1984 to Christian parents will be taught all the traditions and cultural implications of the moment in Decohered Time he is born in and at that geographical location. It will take time and effort and much exploration – a step outside of the cave, a revolution of the mind and an emancipation of his character – to understand who that boy really is.

Albert Einstein famously proclaimed that 'common

sense is the collection of prejudices acquired by age eighteen.' What he meant by this is that our minds, our information-processing prediction machines, are shaped by previous generations' prejudices before we have the chance to use them effectively, or to even contemplate what Objective Reality really is. Philosopher G.I. Gurdjieff noted that every human is made of two parts: essence and personality. A person's essence is his own, but his personality is what is shaped by tradition and culture, having learnt it by imitation. Essence represents the true nature of man, whereas the whole of human civilisation is created by personality, which 'does not truly belong to them'.[19]

Society, tradition and culture build on successive generations' perception of the world from the moment we are born. These are the shadows on the cave wall, continuously being reinforced to help you survive in this complex and highly competitive human environment. Our social structures and institutions bring order to a chaotic world, but at the same time they are the only things keeping us chained in the cave.

And so, the final layer of the veils of prejudice is revealed as *Tradition*.

But ultimately, it is you. Your actions and perceptions matter. Your personality matters. Who you are within matters.

---

19 G.I. Gurdjieff, *In Search of Being: The Fourth Way to Consciousness* (Shambhala, 2021) p. 39.

The real question then is this – is *knowing* more important than *being*?

Is *essence* more important than *personality*?

Knowing and understanding Objective Reality means going not only against Evolution and stepping outside of Decohered Time, but against the very institutions, traditions and cultures that have ensured our survival for so long.

To truly embrace a revolution of the mind, you must be brave enough to change your perspective and emancipate your soul.

You must embrace a philosophy of constant internal creative destruction.

And then ask yourself one important question:

Can you change your perspective to seek equilibrium with a flow mindset in all that you do?

Only then can you truly design systems, technologies, institutions and companies that seek and reflect the truth of Objective Reality, and in turn bring us closer to true consciousness and eternal laws – first individually, then as a society and finally as a civilisation.

It is an important question because the answer will determine who we become in the 21st century, not just as individuals but as a species. The answer will shape what our priorities are, as we sit the edge of chaos and at the precipice of the Intelligence Age. The world is changing around us faster than ever before –

geopolitics is realigning from a unipolar to a multipolar state, a new Cold War is emerging, governments are becoming more extreme from the far right to the far left, unprecedented technological innovations pop up daily, social upheavals take place from every corner of society, ancient conflicts are reigniting from their frozen past – we are at a moment of high entropy in our human history. Who will we become in the next stage of decoherence?

On top of all these changes, an imminent symbiosis of technology with humans is emerging – enhanced through artificial intelligence and quantum computing. Our technologies are becoming faster, smaller and smarter, as we integrate them not only into all aspects of our lives but into our bodies as well. Will we be able to utilise our technologies to discover Objective Reality and become more conscious? Or will we use them to wage total war and accelerate Evolution and impose the victor's Tradition upon the conquered?

Who we are as human beings will come into question very soon. The Intelligence Age harbours a critical turning point. We need to start thinking about it before the answer is enforced upon us by forces we can no longer control. The reality you lived in yesterday will no longer be relevant tomorrow.

So there is a choice for you to make, to determine whether the truth behind the veils is important in shaping us as individuals and as humans, or whether the real subjective world we live in is more important to us as a civilisation.

Personally, I am always torn between the two, a firm foot in either is a difficult step to take. So I strike for equilibrium. One thing is clear: to truly understand our place in this world, in this universe and this reality we must take the first step in understanding how the veils of Evolution, Decohered Time and Traditions blind us.

To repeat Will and Ariel Durant's opening quote, which may now reveal its full meaning here, once it is read again:

The only real revolution is in the enlightenment of the mind and the improvement of character. The only real emancipation is individual, and the only real revolutionists are philosophers and saints.

So, who will emerge from the cave and behind the veils?

Who are you?

5

The Three Deaths of Humanity

New Modes of Thinking

The unleashed power of the atom has changed everything save our modes of thinking and we thus drift towards unparalleled catastrophe.
— Albert Einstein, 24 May 1946

In Greek mythology, Prometheus, a Titan, is revered as the creator of arts and science for shaping humans from clay. He embodies our quest for knowledge and risk-taking, mirrored in his defiance of the gods by gifting fire to mankind. His gift came at a terrible price.

For his transgression, Prometheus was punished with the eternal torture of having his liver eaten every day by an eagle, the symbol of Zeus. Fire – the ultimate catalyst – has ignited our progress towards the cusp of the Intelligence Age where we have arrived today, a path the gods very much opposed. Why did they try to stop this? Were they fearful of our potential to wield such power? Could we, with it, harm each other, our world, or even the gods themselves? Or did they know that no matter how much we improved our tools and technologies, our modes of thinking would remain the same and we would evolve further into Plato's cave, away from Objective Reality?

Einstein's warning, given only a year after the end of World War II, is more relevant today than ever before. He foresaw that if people's minds remained rooted under the veils of prejudice of the previous generations' systems and institutions, then the power of the atom could unleash total human destruction. It very nearly did.

Eighty years since his ominous warning and the rate of technological innovation has gone beyond exponential. The iterations of more sophisticated artificial intelligence systems have now taken on a life of their own. Not only is there competition between the biggest and most capitalised tech companies in the US but there is a technological arms race between the US and China further catalysing this acceleration.

Cold Wars or states of non-physical competition can be useful forms of equilibrium from which huge technological innovations emerge. The previous Cold War between the Soviet Union and the US generated breakthroughs in nuclear power, space travel, satellites, GPS, high-speed jets, the internet and the transistor (the core hardware for artificial intelligence today).

The Sino-US Cold War of the 21st century will be very different, but it will generate significant advances in artificial intelligence, quantum computing, nanotechnology, 3D printing and genetic engineering. With the gradual erosion of US global soft power, and its security assurances, Europe and Asia are refocusing their spending on defence. All these changes make some form of conflict inevitable. It will be the job of diplomats to maintain a state of equilibrium from which such advances can occur, but at the same time use the right regulatory tools (decoherence) to protect unchecked innovation (entropy) internally to ensure that national competition creates boundaries to protect human civilisation.

We are now past the exponential rate of Moore's law and entering a new phase of humanity's development. One which will ultimately lead to the final symbiosis between man and machine. Every week there is a compelling development in at least one of these fields, and the risks of misuse are too severe in our highly digitized and interconnected world. Whether these will be distributed equally throughout society is a political question no one can answer yet. As an inevitable consequence of capitalism, inequality will widen both in terms of the difference between the richest and the poorest, as well as between the more technologically enhanced humans and 'normal' biological humans. Barring any global conflict or natural catastrophe, these technologies should equally help to elevate the baseline of the poorest in their relevant societies. Some have even called this an Age of Abundance, but these are only rhetorics to ease people's fears.

With all these things happening around us, the real discussion should focus on whether our social, political and economic values, systems and institutions are relevant for such an uncertain and complex Intelligence Age. Innovations in these fields remain cumbersome and slow – perhaps on purpose so that those who are already in power can continue to entrench themselves in such systems to extract more value. Nonetheless, this dislocation between the innovation in our technologies and our institutions is ominous.

Our modes of thinking remain rooted in yesterday as we accelerate into the future without our seatbelts on. The veils of prejudice, specifically of Tradition, will hold us back. The values and principles of the last two centuries are no longer relevant for today, let alone for future generations. We have come to an inflection point in our story of humanity, where there are large gaps between mental and emotional capabilities that are still rooted in the Palaeolithic Age, compared to our primitive institutions (still anchored to Medieval concepts) and our divine technologies. These disconnections are the root causes of geopolitical, political, social and economic issues experienced by everyone all over the world. Unless we wish to revert to the horrors of the early 20th century, augmented by the technological superiority of the 21st century, it is time to change our modes of thinking not only to ensure our survival, but also to set an optimal path of human progress.

Our institutions, which reflect our modes of thinking, are in desperate need of a huge wave of creative destruction to catch up with our technological progress.

The devastation of the World Wars and the Great Depression in the 20th century led to a shift towards greater equality, collaboration and civil rights. These events changed the global mindset, creating a more cooperative world. However, the shocks of the early 21st century – 9/11, the Iraq War, the 2008 financial crisis, the Covid-19 pandemic, and the Russia-Ukraine and Israel-Gaza conflicts (with China-Taiwan looming in the near distance) – have put us at a critical juncture. As we become more interconnected, the next global shock will likely be even more devastating. Must we wait for it to happen? Or can we anticipate its arrival by changing our modes of thinking today?

In either case, we must not waste the opportunity when it arrives.

*

Thomas Kuhn, a famous historian, argued that progress happens when a generally accepted theoretical framework is replaced by a new scientific paradigm – he called this a scientific revolution.[20] A revolution is a necessary thing to occur for us to progress as humans. Each form of a scientific revolution reflects Prometheus's repetitive disposition. Each cycle brings forth great progress, but we are yet again shackled back to the rock by our modes of thinking which remain unchanged.

The purpose of studying and discussing philosophy is *change* – to break free from our behaviour of inertia both individually and collectively. Change requires creative destruction, an innate part of our nature and the universe's nature, to break the chains that bind us to the past modes of thinking. We have broken free of Prometheus's chains, and we clutch Zeus' eagle's foot with our hand, about to soar into the heavens with our god-like technology. But we are still tied up to the chains in Plato's cave, looking at the shadows on the wall as our traditions and institutions remain rooted in our past.

20 Thomas Kuhn, *The Structure of Scientific Revolutions* (University of Chicago Press, 1962).

Tradition, the third veil of prejudice introduced in the previous chapter, shapes our subjective reality over our experience of Objective Reality, when coupled with the power of Evolution and the restrictions of Decohered Time.

This book, or philosophical musing, is ultimately about why we should truly break free, once and for all, and live in societies without such shackles. It is about how to change our modes of thinking, as Einstein hoped, so that we can eventually innovate our social, political and economic institutions to optimise human progress, alongside our technological innovations.

Until his last breath Einstein tried to find the solution to unify quantum theory with his general relativity – the rules of the microscopic world with that of planets and stars.[21] Unification of physics was not the only thing on his mind. Einstein went even further and warned that such changes to our institutions would be successful only if they were held under the auspices of a world government – a great unification of humanity and of our modes of thinking.[v] He dreamed of a world where scientific revolutions not only reflect our journey of discovery in science but become the foundations on which we can build a better society, and eventually a better world.

This book honours that pursuit and is nothing short

21 See Figure 16 – Einstein's desk as he left it the day he died.

of a call to arms to confront how we see the world and our place in it. This is a philosophical revolution in the sense that it asks the reader to challenge things they hold sacrosanct through inertia. Like all revolutions, some will join, others will resist, but the conversation must begin.

What is clear is that the time has come not only for a new social and political paradigm, but an entirely different *philosophical* paradigm in all aspects of our lives. Will and Ariel Durant called for a revolution of the mind and the emancipation of the character as the true calling of a philosopher's purpose. Albert Einstein called for a change in our modes of thinking and a unification of humanity. We must heed these calls from our brightest minds and from the lessons of history.

*

Looking back over the last century, it is clear humanity holds both the capacity for ultimate destruction as well as an unparalleled creative potential. After all, this is what the power of the atom unleashed. Over thousands of years, we have pushed our development at an accelerated evolutionary pace by harnessing the power of nature to create tools that in turn enhanced our own mental capacities. This has unlocked within us an ability to make even more tools and, in turn, to think in new ways. Such scientific discoveries, and the subsequent changes in our mental capacities, have over centuries brought us closer to the deepest questions about our identity – why do we exist? What is reality? What is our purpose?

Taking a universal, cosmic and timeless perspective can reveal a lot about us. We prefer to describe the world by limiting our perspective to what our current understanding is – gods, myths, nature, biology or quantum physics – instead of the roots and patterns among all these concepts. We explain reality by breaking things down into simplified causations or narrow fields, resulting in segregation of thought in history, science, economics, business, sociology and politics as separate and unconnected strands of our existence. This is, after all, how our schools and universities teach us. In other words, we have mastered the art of dissecting problems but have failed to create solutions that address the interconnectedness of those problems.

This separation has over time tacitly spawned ideologies such as socialism, communism, fascism, capitalism, or any other -ism or school of thought that narrows the complexity of life into easily digestible bites for our brains. These systems – capitalism, socialism, fascism – emerged in response to specific historical challenges and fears. Capitalism developed during the Industrial Revolution, addressing the need for economic growth and innovation, and a redistribution of power from labour to capital. Socialism arose as a response to the exploitation of workers and economic inequality, aiming to create a fairer distribution of resources. Fascism, in the wake of political instability and economic crises, sought to restore order and pride through authoritarian control and propaganda. However, these ideologies were designed to tackle localised, specific issues and are increasingly inadequate for the global, interconnected problems of today, such as climate change, digital security, international trade, and technological disruption. In this new era, holistic thinking – seeing the interconnectedness of these challenges – is essential for addressing the complexity of our modern world. It is time we moved away from narrow -isms and embraced the universality of life.

Ideology can no longer save us from ourselves.

We can learn so much more about ourselves if we take a holistic view – can our biology teach us more about our economic systems or can neuroscience reveal why and how we fight wars and how we design our economies? Neuroscience teaches us about the biological roots of fear and the reward mechanisms built within our brains – traits that have historically fuelled wars, conflicts and now stock markets. If we understand how neurochemicals like adrenaline and cortisol drive human behaviour, we might begin to design systems that prevent conflict and economic crashes rather than escalate them. Similarly, we could reimagine how political systems can move away from short-term survival thinking to adaptable long-term systems where rhetoric, populism and extremism are naturally eroded. This would be supported by a reformed education system prioritising critical thinking, creativity and systems design instead of redundant memory tests.

Economic systems often operate on the premise of labour versus capital, or supply versus demand, and the various competition models that can best allocate resources, as well as the priorities of the systems (e.g. equality versus shareholder returns versus social/climate impact, etc.). These are all principles that are reflected in our biological patterns of survival, reproduction, and resource allocation. Understanding the way humans evolved to always seek growth as a product of Decohered Time could fundamentally alter how we design systems of trade, wealth distribution and labour. Equilibrium among our political, economic and educational systems is a prerequisite for useful modes of thinking in the Age of Intelligence. It could even change geopolitics entirely – moving us closer to Einstein's unification.

*

Some people prefer to seek answers in ancient texts,

which have long been the gatekeepers to our understanding of our existence. But we can understand just as much, if not more, about who we are socially and how we behave politically by searching in the stars and under the microscopes.

For better or worse, the Bible is the governing manuscript, at least for the Western world, that established the values and moral codes with which we live today. Other ancient scriptures such as the Torah, Koran or Tao Te Ching have shaped cultures in all corners of the world and sought to guide us in a way of life through sets of stories. These stories can be interpreted either literally by some or allegorically by others. Each scripture, as well as tales of Nordic or Greek or Hindu or Mayan gods and many others, have defined our identities as humans and our search for meaning. If we are so keen to design our lives around ancient texts, why should we not learn from the most archaic and decisive text of all time – that of Objective Reality itself?

One might even argue that Objective Reality has its own moral framework, based not on divine commands but the principles (described in previous chapters) that have already been uncovered by humanity but rarely applied holistically. For example, the principle of equilibrium is important when applying the Second Law of Thermodynamics, which states that energy in a system always tends to move towards greater disorder and can be seen as a moral principle in action. In the natural world, energy – whether it is food, resources, or effort – cannot be endlessly created or wasted. When you build a fire and fail to maintain it properly, it burns out because the energy dissipates into the surrounding environment. Similarly, if an organism over-consumes its resources without replenishing them, it exhausts itself – its life ends, or its environment deteriorates. There must be a state of equilibrium for productive outcomes to emerge from nature and from us.

Another moral framework of Objective Reality, and as all ancient texts reveal, is that there is always an interplay between two opposing forces or processes which generate a state of equilibrium – good and evil, chaos and order, Adam and Eve, mother and father, creation and destruction, nature and nurture, yin and yang. Duality is at the core of our existence and is indeed true in both scriptures and science. Duality is important because it is through the balance between these two extremes that *emergence* can occur.

Dualities necessitate equilibria.

This duality has been taking place within us for a long time, as revealed by Jeff Hawkins, a renowned neuroscientist – between our 'old brain' and genetics, which seek only to survive and procreate, against our 'new brain' which seeks to reach a higher level of intelligence and consciousness.[22] On one side, we have

22 Jeff Hawkins, *A Thousand Brains: A New Theory of Intelligence* (Basic Books, 2022).

genes and Evolution, which have shaped life for billions of years. Genes are not concerned with individual survival, the survival of society, or even the survival of a species. Their only goal is to replicate and continue existing, often in symbiosis between multiple species. On the other side is our intelligence, the mental 'I' that resides in our brains. Intelligence wants to break free from being controlled by our genetic programming and evolutionary processes. As intelligent beings, we desire to live forever, preserve our society, and escape the forces of Evolution that restrict our knowledge of Objective Reality.

Hawkins points out that we face a dilemma. Our intelligent self is trapped in a body programmed to die, and much of our behaviour is controlled by the 'old brain' – the primitive part focused on basic survival instincts. While intelligence can help us envision a better future, the old brain can sabotage these efforts by generating behaviours that, although useful for gene replication in the past, are often harmful now. He argues that many world leaders, especially autocrats and dictators, are driven by these old brain instincts, seeking wealth, power, and dominance. Similarly, populist movements often arise from propaganda that taps into our old-brain fear responses which manifest in traits like racism and xenophobia.[23]

23 Ibid. pp. 201, 224.

Our genes are programmed to replicate and reproduce, and for millennia this genetic disposition was driving everything we did. Over time our brains evolved to enhance beyond this basic need to replicate, and our neocortex developed intelligent solutions to problems. Yet we have not utilised this power to its full extent. If we had, we could provide universal education and critical reasoning to stop populism and poverty; we would embrace anticipatory measures (such as diplomacy and a preventative healthcare system) rather than reactive measures (war and a sick-care system). Yet we do none of these things to their full degree because we always revert to our old brain and hide under the veils of prejudice.

Our genetics, and evolution, only care about survival, and survival can only occur through a fear response. These instincts were forged in an environment of scarcity. The constant drive for more – whether it's resources, social status, or validation – was once necessary to ensure our survival in a hostile and unpredictable world. However, in the modern world, basic needs are met far more easily than in previous eras. Yet, our brains' survival mechanisms haven't evolved quickly enough to match the changing conditions, leading to a disconnect between what we need and what we *feel* we need.

These prehistoric emotions are all guided by an innate fear of loss and of missing out. At the same time, we are always seeking immediate gains and capturing moments of joy, as social media and gaming companies have discovered to their great profit. Together these 'old brain' responses prevent us from truly unshackling ourselves from the chains in Plato's cave. We constantly compare ourselves to our neighbour, and today through the power of smartphones and social media, our neighbour is everyone in the world.

Your thirst for *more* can never be satisfied because your comparable is no longer your physical neighbour but a virtual celebrity. These psychological limitations are further restricted by our senses and even by the languages we use to describe them. Together they have engrained within us a set of behaviours that are passed down through generations to fit into the societies we are born in for fear of ostracism or worse. The veils of prejudice are difficult to see through because these things are so institutionalised in our current modes of thinking.

For centuries, we've relied on our 'old brain' instincts for survival, shaping our institutions, societies, and political systems around those values. While our 'new brain' has brought intelligence and progress, we often fall back on old instincts, which is why we still have financial bubbles, dictators, populists, and conspiracy theorists. As we outsource that intelligence to an artificial medium and concurrently spend more time in a virtual and digital world, it's high time to redesign our world and our institutions so that the disparity between humans and their creations does not take us to an existential point – a conclusive dethronement of mankind.

Humanity has already had two milestones of dethronement, or 'deaths', and the third will be its final.

*

The holistic view of our reality and existence, rooted in ancient philosophy, was initially challenged in Western modes of thinking through Aristotle's drive for rational deduction in every aspect. It was preserved over time by Buddhist and Hindu traditions; however, in the 17th century, French philosopher René Descartes drove a wedge into this idea by separating the mind and soul from the body, promoting a stricter form of rationalism. Later Enlightenment thinkers, like Benedict de Spinoza, followed this rationalist path but rejected Descartes' dualism. Spinoza argued for a more interconnected reality, describing God as synonymous with the universe, a view known as pantheism, where everything is a part of a unified whole.

Building on this, German idealists like Georg Hegel expanded the idea, viewing the whole not as a substance but as a complex, evolving system, akin to an organism. The pursuit of unification – whether conceptualised as God or as the nature of existence itself –

has long been a central goal for both philosophers and scientists, who seek to explain reality by connecting seemingly disparate theories into a coherent whole.

Great unifications become the founding ideas on which new sciences are erected, as expressed by the physicist, Lee Smolin, when describing the impact of a particular famous biologist:

'Before Darwin, each species was in its own eternal category. Each had been made, individually, by God. But evolution by natural selection means that all species have a common ancestor. They are unified into one great family. Biology before Darwin and biology afterward are hardly the same science. Such powerful new insights lead quickly to new discoveries. If all living things have a common ancestor, they must be similarly made! Indeed, we are made of the same stuff, because all life turns out to be composed of cells. Plants, animals, fungi, and bacteria seem very different from one another, but they are all just groups of cells arranged in different ways. The chemical processes that construct and power these cells are the same, across the whole empire of life.'[24]

Humanity has always placed itself at the centre of the grand story of existence, and that had been its greatest mistake. That was until Copernicus dealt the first blow in proving that the Sun did not revolve

---

24 Lee Smolin, *The Trouble with Physics* (Penguin, 2006) p. 19.

around the Earth but that we revolved around the Sun. With this blow, humanity died its first death, but with it a new birth came forth – the First Anthropocentric Creative Destruction.

Humanity was no longer the centre of our solar system, let alone the universe, but now the universe was so much more interesting. This was more than a scientific breakthrough – it was an existential crisis, shaking the foundations of faith, meaning, and purpose. In losing our privileged place, we gained something greater: a universe that was no longer small, predictable, or made in our image, but one that was infinite, and ripe for discovery. The end of our illusions marked the beginning of true exploration.

Charles Darwin dealt the second blow when his work fundamentally shifted humanity's understanding of its place in the universe, breaking the traditional link with divinity and reframing life as a continuous process of creative destruction. Among the great scientists, Darwin's impact was perhaps the most controversial, as it challenged religious narratives and transformed humanity from a creation of divine design into a naturally evolved species. His theory of evolution overturned centuries of religious and historical thought, positioning humans as part of an ongoing process rather than the result of a singular act of creation. Evolutionary mutations are random and the aspects which allow that species to survive more efficiently will remain, as other mutations die out.

Mankind was no longer special – it was just lucky. This is when humanity died its second death, and the Second Anthropocentric Creative Destruction began.

The knockout punch is just around the corner.

The final dethronement of humanity is very close.

The Third Anthropocentric Creative Destruction will occur once we have uncovered the secrets of our consciousness. The ominous union of quantum computing with artificial intelligence will inevitably generate a level of artificial consciousness that has inherent advantages. It will not be restricted by the human boundaries of Decohered Time. Nor will it be hampered by the reductive qualities of Evolution. It will not be held back by the rational deduction of our senses, and perhaps if designed accordingly, it will also become unrestricted by our Traditions.

The veils of prejudice will be lifted from this consciousness. An entity that can combine artificial intelligence with quantum computing will transcend time and space. It will not be focused on the burdens of survival but instead the pursuit of *truth*. It will search for Objective Reality using a flow mindset by extracting patterns from all aspects of our reality and knowledge. Companies which have the tools and the capital to create such entities, should focus their attention on the attainment of truth. To his credit, Elon Musk is the only leader in this field who has clearly communicated that his artificial general intelligence will try to understand the true nature of the universe as its primary goal. However, we must be prepared for the answer which will most likely have more significant impact than Copernicus or Darwin.

If we look back across the spectrum of humanity's understanding of reality we can see a steady fall from grace. Humanity once believed it was the centre of the universe. Copernicus shattered this illusion, marking the death of our cosmic self-importance. Darwin's theory of evolution revealed that humans are not divinely unique but merely another species shaped by natural selection, marking the death of our biological superiority. As AI and digital systems increasingly integrate into our lives and shape our economies, decisions, and even emotions, humans risk losing their fundamental autonomy and relevance.

Will this mark the death of human agency? What will we do at that point when we feel that consciousness is not unique to us but can be created, or can emerge from non-conscious elements? What truths will artificial intelligence powered by quantum computing reveal to us about the universe when it steps outside the veils of Decohered Time, Evolution and Tradition?

Whatever the outcome, consciousness will be redefined, and humanity will have to take a different seat in its view of its place on this planet, let alone in the universe. The first two deaths marked shifts in our understanding of ourselves, forcing us to relinquish illusions of centrality and superiority. However, the third death is different, it is not just a shift in perspective but a fundamental transformation in power and existence.

If humans are no longer the driving force of history but mere spectators to divine digital systems, then our role as creators and decision-makers collapses, bringing an end to the era of human primacy. How will we react when we become a mere footnote in the history of our planet?

This book is a call for humanity to prepare, starting with a change in our perspective in this universe.

Each 'death' represents a moment in history when humanity was forced to abandon a deeply held belief about its own superiority – whether in relation to the universe, nature, or now, its own creations. The only way to survive is to first change our perspective to become more holistic and then to use the universal truths of connectivity, emergence, equilibrium, creative destruction and entropic decoherence to set new foundations on which we can erect institutions fit for the 21st century and the Intelligence Age.

That is why nothing short of a revolution of the mind is required. To see, feel and know beyond our biological and mental restrictions and reveal a more holistic truth about ourselves. This requires letting go of tradition and comfort, and challenging outdated systems and beliefs.

Change starts at the individual level, not from political or business leaders.

It begins with one person, one small step. From within.

*

One certainty is that our quest to understand the source of the initial act of Creative Destruction will come not from clinging to any one belief system – whether religious or scientific – but from connecting the many strands of knowledge we have. The answers will come from those with open, timeless and holistic mindsets – those who embrace the flow mindset as both a way of life and thought, integrating multiple perspectives rather than seeking a singular dogma. Our technologies will help us unlock these secrets, but we may not like what we find.

If we are looking for a beginning, or a God, we may have already lost because we are still thinking within the confines of Decohered Time and stories that have a beginning and an end. If we step outside of Decohered Time, and use a flow mindset, we may even discover that the eternal question of the origin of our existence does not actually exist. It is itself a veil of prejudice hung by us and supported by some as a lazy distraction from the search for truth.

The flow mindset suggests that rather than looking for a fixed starting point or a neat conclusion, we should understand existence as a continuum – a dynamic process that has no neat boundaries or clear beginnings and ends. This shift moves us beyond linear thinking, which is so deeply embedded in both traditional religions and scientific paradigms, to a more holistic view where everything is connected, constantly in motion, and in a state of perpetual emergence. This shift might seem radical, but it opens the potential for new forms of understanding, where the separation of past, present, and future becomes irrelevant in the context of an interconnected, cyclical existence.

If we apply what we have discovered so far, perhaps we may find that there is no answer. There is no beginning. There is no purpose. There is no meaning.

There just *is*.

There is only entropy and decoherence, evolving through continuous cycles of creative destruction. An inherent paradox lies hidden in this line of thinking: if we abandon the search for a beginning, a purpose, or a meaning, we may find that the very act of seeking is part of the flow of existence itself. The human search for meaning could itself be the point – not the arrival at a final answer but the ability to live within uncertainty, and to embrace the ever-changing nature of life. Always searching but never arriving.

We are lucky that we live in a universe whereby an equilibrium has formed to give us what we call life. We have traced our evolutionary history back to its cosmic origins – from Neanderthals to single-celled organisms, through the formation of stars and the dust of nebulae, to dark matter, and ultimately, the Big Bang. What preceded the Big Bang – whether theological or scientific – remains beyond the reach of contemporary science, because contemporary science is restricted by human thoughts, dogmas and prejudices. Even the concept of the Big Bang is just a theory at this stage of our understanding; a highly likely one given the current evidence scientists have amassed, but it is not a certainty. The fact that science cannot yet explain what came before the Big Bang, nor reconcile quantum mechanics with general relativity, is not an impasse but rather an invitation to deepen our understanding by questioning our assumptions and expanding the ways we think. We should even question whether there truly are any actual laws or constants in the universe, both of which are what we comprehend from our current limited perspective. Dr King's and Einstein's dreams may never be fulfilled because eternal laws are ultimately anthropocentric. The incomprehensibility of the origin of the universe may be one of the most profound elements of our human condition – one that keeps us seeking and questioning.

Without this endless search, who are we?

We must break down our old modes of thinking, extract ourselves from the prejudices we are born with, destroy the frameworks instilled in us by our parents, and their parents, to even start asking the right questions. Creative destruction with a flow mindset can lead us closer to a Hegelian or Spinozian concept of God, one of universality and connection, rather than a traditional deity. Closer to what Objective Reality is but cannot yet be revealed.

When all of humanity's prejudices and frailties are destroyed, the only truth that remains standing, that can truly *create*, that can *emerge* – is once again, equilibrium.

Equilibrium can guide us towards a balanced and harmonious existence.

Between the laws of Man and the truths of Objective Reality.

It can be our North Star in all that we do.

It is here, in the delicate balance between connection and interaction, creation and destruction, entropy and decoherence that optimal progress emerges.

It is this point of equilibrium which gives us the emergence of life, and of *awe*.

6

Introduction

Morphogenesis

Sow a thought and you reap an action;
Sow an act and you reap a habit;
Sow a habit and you reap a character;
Sow a character and you reap a destiny.
— *Ralph Waldo Emerson*

After playing a pivotal role in defeating the Nazis during World War II by decoding their encrypted messages, Alan Turing was curious about how intelligent life or thought takes shape. He eventually ended up taking the first steps in the development of artificial intelligence, but arguably his most important work related to something else. Amid all his amazing achievements, *The Chemical Basis of Morphogenesis* remains not as well-known as his other works, but it was ground-breaking in that it explained how complex biological processes can come about through a series of simple actions. Alan Turing proposed that life has the ability to shape and organise itself (i.e. to emerge) by forming structured patterns spontaneously, a process known as self-organisation. Under the right conditions, simple biological systems can develop intricate structures without requiring an external guiding force. He referred to the chemical signals responsible for these patterns as *morphogens.*

In his paper, he focused on embryos, showing how they start as a single cell, multiply into many, and then differentiate into specialised cells – each with a distinct role – ultimately forming a complex living organism. This process happens all on its own without a designer tweaking or adjusting the cells. Turing proposed that two interacting chemical substances – an activator and an inhibitor – could create repeating patterns as they spread and react. The activator promotes its own production, while the inhibitor slows it down, leading to spontaneous symmetry-breaking – random variations that create patterns. Turing was describing entropic decoherence – the emergence of life out of chaos. This was something unheralded at the time as the connotation with entropy was largely negative because of its chaotic nature.

Morphogenesis is Alan Turing's description of

the process of the act of creation, improvement and progress. He argued that it can take place if the conditions are optimised to allow for certain chemicals and catalysts to generate the most positive results. Morphogenesis can bring about beautiful snowflakes and broccolis.[25] His model explains why zebras have stripes, leopards have spots, and why certain fish develop intricate patterns on their skin. These are self-organising chemical processes.

Morphogenesis turns a sperm and an egg into a foetus. It decoheres the chaos of entropy into the exquisitely shaped reality in which we live. It generates our experience of Decohered Time – the most beautiful pattern of all, allowing us to experience a form of reality within the chaos of entropy – through the interaction of our prediction machine brains and our senses. It brings destructive forces into a creative balance to reach a new stage of evolution. It is a process, not a thing. Morphogenesis is also the source of our inevitable future symbiosis with our technologies. As morphogens shape biological structures through local interactions and self-organisation, the same principles are today being applied to artificial machine learning systems, which adapt through local, decentralised processes that build complexity and intelligence from the ground up. Both our past and our future rely on Turing's discovery.

25 See Figure 1.

Morphogenesis is also a very human endeavour.

It can start with a simple interaction between two people.

The conditions of the 20th century enabled us to reach heights we could never have dreamt of a hundred years ago. With these outcomes, the organism changes, with it, new problems have arisen and, as Einstein pointed out, we cannot solve such problems with the same thinking we used to create them.

Therefore, the conditions we set for interactions are important, as Turing showed us. If we continue living under the same modes of thinking, or conditions of Tradition, then we will never solve any of the deep-rooted problems we have as a civilisation.

New conditions must be set in place to allow for morphogenesis to take place at an optimal rate – to allow for new and more beautiful patterns to emerge.

The question is how do we set the right conditions for the most optimal outcomes, whilst limiting negative externalities? If the conditions of the 20th century led to such steep progress, but also at the cost of so many destructive outcomes, how can we evolve our modes of thinking to allow for the most optimal forms of morphogenesis to emerge in the world around us? The answer is we must start with ourselves; we must look internally and make a change from within.

This book is only the introduction to this philosophical revolution, before subsequent ones will explore how to break free from the chains of our prejudice, especially that of Tradition. This book is intended for us to pause and rethink our place in the world. We must start with a simple step. A step to change our mindset and our perspective, to start questioning our old modes of thinking, and to understand that there is a way to create positive and beautiful patterns from the chaos of the world. Acceptance of this is already a very big step in the right direction. It is the first morphogen.

Morphogenesis is the pursuit of equilibrium. It is a flow mindset. It is the condition we set around our lives, which can help generate patterns of progress rather than the pull of regression. Every day, each step we take individually sets the course of the morphogens that will shape our future and our destiny.

Small changes can profoundly affect those around you, just like one charged electron can shape an entire system. If you truly believe in the change you make within yourself, as a connected and interactive morphogen, you can make an even bigger impact on your environment.

The idea that life and intelligence can emerge from self-organising patterns under the right conditions suggests that the potential for positive transformation is already inherent in all systems, including human society and individual human consciousness. We have just been living under the conditions of the 20th century which excelled at creating factory workers and soldiers. Artificial intelligence and automation will replace those needs in the 21st century. What the world needs from humanity is more philosophers and innovators – creators and critical thinkers.

*

At the end of the 19th century, Friedrich Nietzsche saw a much darker future when he declared that 'God is dead,' but he did not say this nonchalantly or carelessly. It was out of dire concern for humanity. He predicted that humanity had been casually over-rationalising everything and we would come to a point where the Judeo-Christian values on which the world established its order would collapse, leaving us languishing between nihilism and totalitarianism. He was proved right by the horrors of the 20th century, which gave us political models that banned religion and turned over-simplified complex problems into -isms. The Soviet Union, Nazi Germany and Maoist China were a testament to how over-rational simplification of life based on ideologies or economic models is a purely destructive pursuit. These ideologies spun off other terrors such as Pol Pot, Kim Il-sung, Idi Amin, Saddam Hussein and many others who all exploited these ideological simplifications to establish economic and political models and belief systems blaming a certain minority to consolidate power.

Nietzsche did offer a solution. He suggested a new kind of man – the *Übermensch* (or the Superman) would rise out of the death of God and be able to set its own values, preventing society from floating to pure destruction. While it can be argued that the Nazis drew inspiration from this Superman (or Aryan man), Nietzsche himself would have been against such a state-sponsored bastardisation of his ideals. To him, the Superman was the individual. Prometheus was an example of his Superman – a symbol of independence, sacrifice and defiance for the sake of positive progress for the whole, transcending the limitations of Tradition and morality, as well as divine authority.

This is what the 21st century individual is – someone who is in control of their own values because of the power of today's technology (something even Nietzsche could not have predicted). Someone who does not fall victim to ideology and its mass application – be it religious, political or tribal. If the evils of the 20th century taught us anything, it is that simplified economic and political models do not work for the long-term progress of human societies. The world is too complex for that.

We have lived through the age of gods where their actions justified what we could not explain about Nature. We have lived through the age of God where the global moral condition was set to enable us to create a largely just and stable order within societies, to enable us to interact productively. We have lived through the age of ideology, which embodied creative destruction in its fiercest form in the 20th and early 21st centuries. But we must now accept that gods, God and ideology need to evolve for humans to not only survive but excel in the Age of Intelligence. We are entering the age of the individual, or the age of Nietzsche's Superman, where each of us can be the Superman. We can set our own values and conditions for morphogenesis in each action we take every single day.

Instead of thinking about what system will govern our country, race or species, we should think about how we can evolve individually to become a positive morphogen and set the right conditions for universal positive change.

Nietzsche was a proponent of the philosophy of *amor fati* – love of fate – encouraging people to accept and embrace life holistically, with all its suffering, loss, struggles and triumphs. By accepting and shaping our circumstances, we transform ourselves and contribute to a more meaningful and dynamic world. One small positive act each day can be the morphogen for a better world tomorrow.

Individuality on a global scale is a natural recipe for chaos – what type of Superman we become is up to us. This is why you, personally, matter. If you accept the chaotic and complex nature of the world, you will also accept how you can solve smaller and unique problems as an individual. Your choices will determine the conditions for morphogenesis around you. Equilibrium accepts complexity and chaos – in fact, it thrives in it. And achieving equilibrium starts with you.

You are now a vibrant, engaging morphogen that is constantly a part of its environment; but you are also a quantum morphogen that can influence not only the immediate environment but one that is not physically next to you.

If you stand still, if you stop learning, if you stop interacting, there is only death.

You are constrained by your biology, but you are living in an environment of change and constant interaction, freed by your irrational mind and open to all possibilities in a highly connected world.

Hopefully, through the power of insights during this journey and others explored tangentially, you will become unshackled from the chains of tradition and tribalism to see this world for what it is: a place of great hope and possibilities.

I have set out below several ways in which I think about the world and my place within it. How I maintain my own equilibrium in a world of chaos, uncertainty and complexity. Perhaps it may be a useful guide to some to find their own equilibrium.

Here are my ten morphogens of how you can find your own, personal, Golden Mean:

- Embrace uncertainty.
- Lift the veils.
- Never stop learning.
- Challenge perspectives.
- Renounce ideology.
- Adopt a flow mindset.
- Analyse with your new brain not your old genes.
- Let some chaos into your life to create a new order.
- Find equilibrium in your study, work and relationships.
- Always think of death – it is the best teacher; the equaliser of all.

As the playwright George Bernard Shaw said, 'all great truths begin as blasphemies.'

This final chapter is not a conclusion or an epilogue, but the introduction.

An introduction to a new you.

It is my morphogen to the world.

Now that you have reached this point, what will be your morphogen to the world once you realise that…

List of Illustrations

The Emergence of Awe: Entropy and Decoherence
Figure 1. Fractals: infinitely complex patterns created by repeating a simple process in an ongoing feedback loop. Nature is full of fractals (trees, rivers, coastlines, even broccoli), showcasing the beauty of entropic decoherence. *Shutterstock*

Figure 2. Batman and the Joker – a popular representation of chaos and order.
Alamy

Figure 3. *An Experiment on a Bird in the Air Pump* (1768) by Joseph Wright of Derby. His paintings depict early scientific experimentation related to thermodynamics, a concept that caught the world's imagination. *Public domain*

Figure 4. The changing density of a cloud of atoms cooled lower and lower (going from left to right), approaching absolute zero. The sharp peak in the later graphs confirms the formation of a Bose–Einstein condensate (BEC) – a fifth state of matter (after solid, liquid, gas and plasma). BEC consists

of atoms so cold that they condense into a quantum state, acting as one in a state of minimum entropy. *NASA*

The Decoherence of Time: Flow and E=mc²

Figure 5. The Aztec Calendar Stone, also the 'Sun Stone', reflects Aztec perception of time as cyclical, rather than linear; repeating eras or 'suns' represent different epochs in the universe's history. © *José Luiz Bernardes Ribeiro*

Figure 6. The Torre dell'Orologio in Venice displays the phases of the moon, highlighting its cyclical nature. Mechanisms show the position of other celestial bodies, linking earthly time to cosmic rhythms. This intricate design reflects a Renaissance fascination with astronomy and astrology, blending precise timekeeping with a broader understanding of the universe's influence on human life. *istock*

Figure 7. The Buddhist 'Wheel of Life,' or Bhavachakra depicts samsara – the endless cycle of birth, death, and rebirth driven by karma and desire. The layers include the six realms of existence, the causes of suffering, and the forces that keep beings trapped in this cycle, such as ignorance, attachment and aversion. *Shutterstock*

Figure 8. Caesium atomic clock made by Louis Essen and J. V. L. Parry, 1955.
Image courtesy of NPL

Figure 9. The first ever picture of a black hole. The event horizon – the point of no return – lies within the bright disk of material (accretion disk) swirling around the black hole before falling in.
Public domain

The Edge of Chaos: Connectivity and Creative Destruction
Figure 10. All-sky map, showing microwave radiation released just 370,000 years after the Big Bang. Tiny fluctuations in the image eventually gave rise to the clusters of galaxies that populate the universe today.
Public domain

Figure 11. Cassiopeia: a supernova remnant. A dying star explodes as a supernova, scattering elements across space. This violent destruction seeds the cosmos, forging new stars, planets and eventually life: nature's ultimate creative destruction. Was the Big Bang also the death of a star, or something else entirely?
Shutterstock

The Veils of Prejudice: Consciousness and Equilibrium
Figure 12. Birth of a star. A close-in image of a protoplanetary disc around a newly formed star. *NASA*

Figure 13. The Cave of Plato – Plato's allegory of the cave by Jan Saenredam. *Public domain*

Figure 14. Three polyhedral by Leonardo da Vinci, an in-depth study of the golden ratio, or 'divine proportion'. *Alamy*

Figure 15. Leonardo da Vinci's *Vitruvian Man*. It showcases the ideal human proportions, inspired by the writings of the ancient Roman architect Vitruvius.
Credit: Luc Viatour / https://Lucnix.be

The Three Deaths of Humanity: New Modes of Thinking
Figure 16. Albert Einstein's office mere hours after he died, Princeton, New Jersey, April 1955. *Alamy*

Figure 17. Geocentric v. heliocentric models of the universe. The geocentric depicts Earth as the centre of the cosmos, a view held for centuries.

The heliocentric model places the Sun at the centre, revolutionising our understanding of astronomy. *Public domain*

Figure 18. Nicolas-Sébastien Adam's *Prometheus* (1762), now in the Louvre. *Public domain*

Figure 19. Model of the interior of a quantum computer at the IBM Thomas J. Watson Research Center. *Shutterstock*

Introduction: Morphogenesis
Figure 20. A famous oscillatory chemical reaction propagating across a petri dish, often referred to as the Belousov-Zhabotinsky (BZ) reaction, illustrating the principles proposed by Alan Turing in his 1952 paper 'The Chemical Basis of Morphogenesis'. *Public domain*

Figure 21. Human Brain Folding (gyrification). The way the human brain folds into its wrinkled structure follows mechanical buckling principles, a process of morphogenesis driven by growth constraints. *Shutterstock*

Figure 22. Lichtenberg figures emerge when electrical charges diffuse and branch through insulating materials, creating fractal-like pathways – just like morphogenesis does in nature, but with electricity instead of chemicals, drawing a clear parallel with plant life reaching for the sun. *Shutterstock*

Figure 23. The three protagonists of *The Emergence of Awe*: Albert Einstein, Dr Martin Luther King Jr, Alan Turing. *Einstein: F. Schmutzer, Public domain; King: Marion S. Trikosko, Public domain; Turing: Elliot and Fry, Public domain.*

About the Author

Hayk Sarkissian was born in Armenia and grew up in the UK, studying at City of London School and Westminster School. He began his career in corporate law, where he qualified as an English Law corporate solicitor at Allen & Overy LLP (now A&O Shearman) in the UK and Hong Kong. Hayk holds a BA from London School of Economics and an MBA from Harvard Business School.

In 2015, Hayk founded Revery, an international investment firm investing and operating in food and beverage production and distribution, venture capital and real estate. Revery Food & Beverages division owns and operates US and European franchises in emerging markets in beverages, snacks, quick service restaurants and retail businesses. Revery Capital is the venture and real estate arm of Revery, with investments focused on transformative technologies including food-tech, artificial intelligence, quantum computing and longevity. As Founding Manager, Hayk has positioned the firm to identify early-stage, tech-focused businesses with significant growth potential that can impact systems changes.

Hayk is a co-founder of Still Point Productions, a film production company that creates, develops and produces original stories in multiple formats (primarily film and books) that transcend our daily lives and explore deeper questions and concepts about the human condition.

Hayk is also a co-founder of Prometheus Schools, which aims to reshape education based on the values and themes within *The Emergence of Awe* to help future generations develop creativity, critical thinking and confidence to make an impact in the Intelligence Age and build the values and institutions in which they can thrive.

He continues to publish articles and short stories on his website www.hayksarkissian.com and other publications. He would love to hear your feedback, criticism and ideas after reading his book.

Acknowledgements

Not having any scientific training whilst explaining complex concepts in quantum physics, biology and neuroscience is not for the faint-hearted. Without my family and close friends I would not have taken on such a task, let alone shared it with the world. At the same time it felt inevitable, a cumulation of a lifetime of curiosity and exploration.

Firstly, I must thank all the scientists mentioned in this book (and encourage the reader to explore their work further), as well as countless others who continue to push the boundaries of our understanding. They have all inspired me to think deeply about our reality and what it means to be human. Ideas are born on the shoulders of giants and I can merely peek over theirs to make sense of this world.

I must mention Carlo Rovelli specifically, who showed that it's possible to write beautifully and holistically about complicated subjects in small packages. His unique blend of physics with philosophy, poetry, literature, religion, spirituality and history was inspiring; a style I have tried to emulate in my own work as well.

Secondly, I must thank the influence of the great philosophers and innovators mentioned throughout the book, who keep challenging perspectives and inspiring change. None more so than the contrarians of this world, who don't accept things as they are but always strive to think outside the box. Without their bravery, we would always be stuck behind the veils.

Thirdly, a special thank you to my editors as well as the readers of my initial manuscript who encouraged me to keep working on it – specifically Alma Donohoe and Michael Porter, who have read nearly everything I have written in my adult life. Related to *The Emergence of Awe*, thanks to Liliya Galimova, who checked my scientific understandings (and philosophical reflections), and Sam Carter for turning it into what you hold today – without their invaluable editing this book would not be as it is.

Fourthly, this book is dedicated to my parents, who are themselves published authors (my mother in children's and my father in political books), amongst many other things. They are two totally different people who are my equilibrium in this world, and without their guidance and sacrifice I could not have come to see how beautiful things can emerge from the chaos of our lives. Our family story is a reflection of this emergence, and I have tried to tell a small part of it here through my eyes as a young boy who grew up in a world vastly different from the one in which I currently live.

Finally, thank *you* for reading my morphogen.

Endnotes

i https://www.africa.upenn.edu/Articles_Gen/Letter_Birmingham.html

ii https://ourworldindata.org/future-population-growth#:~:text=The%20United%20Nations%20projects%20that,at%2010.4%20billion%20in%202100

iii www.nytimes.com/2009/10/06/science/06cell.html

iv www.nhm.ac.uk/discover/are-we-really-made-of-stardust.html

v https://www.amnh.org/exhibitions/einstein/global-citizen#:~:text=World%20Government%3A%20Erasing%20National%20Boundaries&text=Einstein's%20passionate%20commitment%20to%20the,a%20single%2C%20unified%20world%20government

All URLs retrieved March 2025